RESILI

Is resilience simply a fad, or is it a new way of thinking about human–environment relations, and the governance of these relations, that has real staying power? Is resilience a dangerous, depoliticizing concept that neuters incipient political activity, or the key to more empowering, emancipatory, and participatory forms of environmental management?

Resilience offers an advanced introduction to these debates. It provides students with a detailed review of how the concept emerged from a small corner of ecology to critically challenge conventional environmental management practices, and radicalize how we can think about and manage social and ecological change. But *Resilience* also situates this new style of thought and management within a particular historical and geographical context. It traces the roots of resilience to the cybernetically influenced behavioral science of Herbert Simon, the neoliberal political economic theory of new institutional economics, the pragmatist philosophy of John Dewey, and the modernist design aesthetic of the Bauhaus school. These diverse roots are what distinguish resilience approaches from other ways of studying human–environment relations. Resilience thinking recalibrates the study of social and environmental change around a *will to design*, a drive or desire to synthesize diverse forms of knowledge and develop collaborative, cross-boundary solutions to complex problems. In contrast to the modes of analysis and critique found in geography and cognate disciplines, resilience approaches strive to pragmatically transform human–environment relations in ways that will produce more sustainable futures for complex social and ecological systems.

In providing a road map to debates over resilience that brings together research from geography, anthropology, sociology, international relations, and philosophy, this book gives readers the conceptual and theoretical tools necessary to engage with political and ethical questions about how we can and should live together in an increasingly interconnected and unpredictable world.

Kevin Grove is Assistant Professor in the Department of Global & Sociocultural Studies at Florida International University, USA.

Key Ideas in Geography

SERIES EDITORS: NOEL CASTREE, UNIVERSITY OF WOLLONGONG AND AUDREY KOBAYASHI, QUEEN'S UNIVERSITY

The *Key Ideas in Geography* series will provide strong, original and accessible texts on important spatial concepts for academics and students working in the fields of geography, sociology and anthropology, as well as the interdisciplinary fields of urban and rural studies, development and cultural studies. Each text will locate a key idea within its traditions of thought, provide grounds for understanding its various usages and meanings and offer critical discussion of the contribution of relevant authors and thinkers.

For a full list of titles in this series, please visit www.routledge.com/series/KIG

Scale
ANDREW HEROD

Rural
MICHAEL WOODS

Citizenship
RICHARD YARWOOD

Wilderness
PHILLIP VANNINI AND APRIL VANNINI

Creativity
HARRIET HAWKINS

Migration, Second Edition
MICHAEL SAMERS AND MICHAEL COLLYER

Mobility, Second Edition
PETER ADEY

City, Second Edition
PHIL HUBBARD

Resilience
KEVIN GROVE

RESILIENCE

Kevin Grove

LONDON AND NEW YORK

First published 2018
by Routledge
2 Park Square, Milton Park, Abingdon, Oxon OX14 4RN

and by Routledge
711 Third Avenue, New York, NY 10017

Routledge is an imprint of the Taylor & Francis Group, an informa business

British Library Cataloguing-in-Publication Data
A catalogue record for this book is available from the British Library

Library of Congress Cataloging-in-Publication Data
A catalog record for this book has been requested

ISBN: 978-1-138-94902-7 (hbk)
ISBN: 978-1-138-94903-4 (pbk)
ISBN: 978-1-315-66140-7 (ebk)

Typeset in Joanna MT Std
by Apex CoVantage, LLC

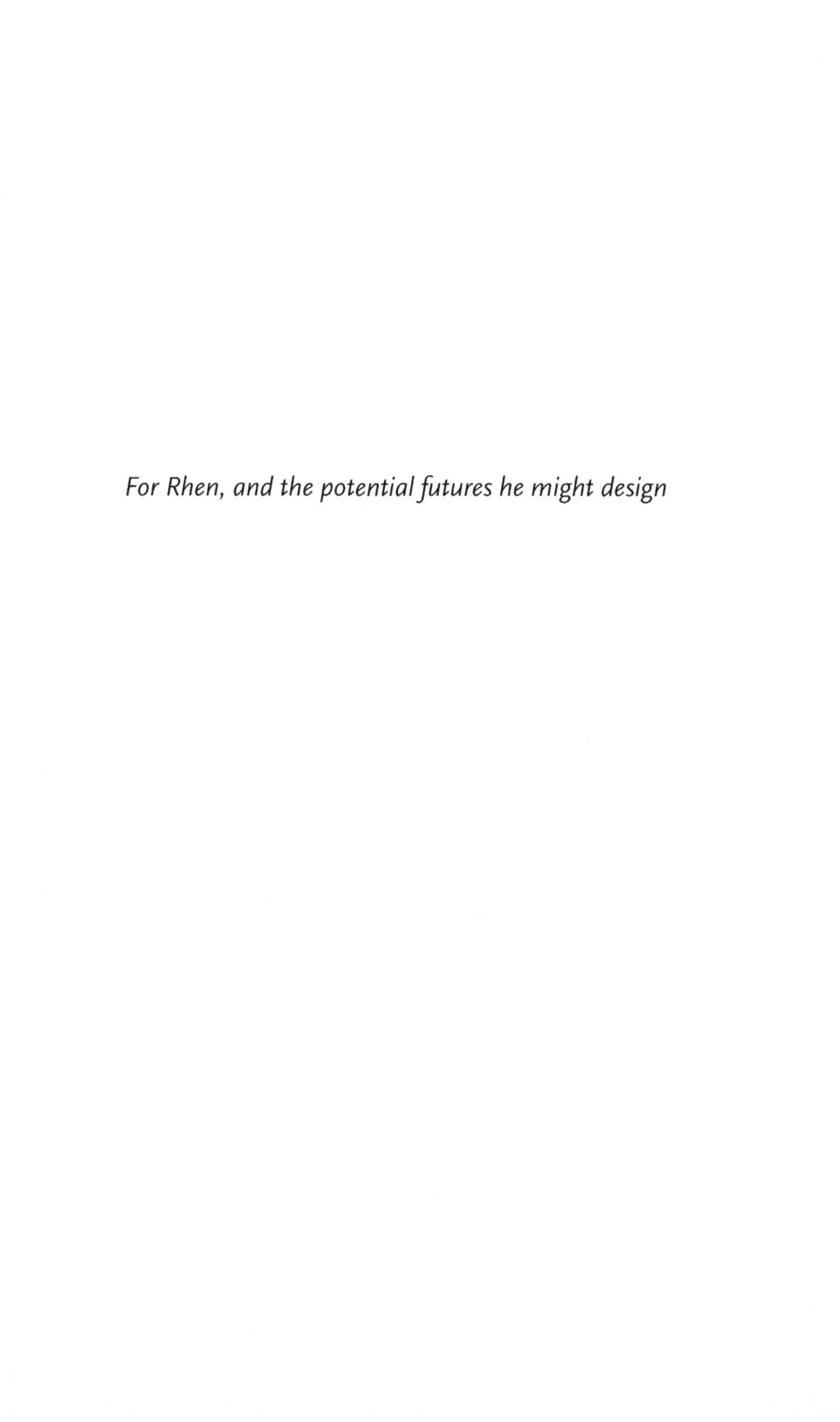

For Rhen, and the potential futures he might design

Contents

ACKNOWLEDGEMENTS ix
LIST OF FIGURES xiii

1 Resilience and geographic thought 1

2 Resilience as essentially contested concept 30

3 Resilience as subjugated knowledge 65

4 Resilience as critique 89

5 Resilience as design 131

6 Resilience as control 182

7 Un-worlding resilience 228

8 Conclusions: re-designing resilience? 265

REFERENCES 277
INDEX 298

Acknowledgements

Any book is an assemblage, and this one is no different. While the words, ideas and arguments are mine, they have been inflected through numerous encounters and exchanges in a variety of places, and I am grateful for all of the productive interactions and supportive encouragement I have received while crafting this work. There are too many people and places to name, but I will try. Ben Anderson warmly invited me to share ideas at "Governing Emergency" workshops in London, New York City, and Adelaide, and the community of scholars that has coalesced through that network has been particularly inspiring to work with. Stephen Collier and Savannah Cox introduced me to the writings of Herbert Simon while we were discussing design-driven resilience in New York, and it was an eye-opening experience. Reading Simon was like reading Holling, Lee, and other key resilience scholars, but written decades earlier. This opened new paths of exploration that would not have otherwise been possible, and I am exceptionally grateful for their inspiration. Lauren Rickards and the Center for Urban Research at the Royal Melbourne Institute of Technology graciously hosted me as a Visiting International Fellow during summer 2016, which gave me a chance to present initial versions of some of the ideas in this book to a variety of audiences. I have particularly valued Lauren's provocative and supportive discussions over email, Skype and in person ever since. David Chandler provided incredibly useful feedback on earlier versions of two chapters at the 2017 Boston AAG and Political Geography Pre-Conference, and our co-edited special issue of *Resilience: International Policies, Pratices and Discourses* gave me a chance to think through many ideas that inform this book's later chapters. I have also cherished the support and encouragement of many other colleagues as I worked through the manuscript. In Miami, Rod Neumann,

Gail Hollander and Arelis Lopez helped create a welcoming and supportive environment at FIU that enabled me to complete this project. Evelyn Gaiser, John Kominoski, Tiffany Troxler, Nancy Grimm and Charles Redman provided me an opportunity to participate in the UREx-SRN, which has been a continual source of academic and pragmatic inspiration for this book and for my current research activities. In Aberystwyth, Kimberley Peters, Jennifer Turner, Mitch Rose, Gareth Hoskins and Mark Whitehead all encouraged the initial stages of this work. Many others have shared their support and feedback along the way, including but certainly not limited to Claudia Aradau, Melissa Bernardo, Jon Coaffee, Mat Coleman, Raven Cretney, Simon Dalby, Mona Domosh, Arturo Escobar, Elizabeth Gagen, Hugh Gladwin, Jesse Healey, Rhys Jones, Rhys Dafydd Jones, Matt Kearnes, Garnet Kindervater Maros Krivy, Lauren Martin, Josh Mullenite, Sara Nelson, Nat O'Grady, Shannon O'Lear, Elspeth Opperman, Julian Reid, David Robles, Stephanie Wakefield, and Mike Woods.

Many ideas and lines of argument presented here have been presented and refined at a variety of talks and presentations. I'm particularly grateful for Jonathan Pugh and audiences at Newcastle University, Andrew Baldwin and audiences at Durham University, Samuel Randalls, Jason Dittmer and audiences at University College London, and Jennifer Lawrence, Simin Davoudi, and the Global Forum on Urban and Regional Resilience at Virginia Tech for opportunities to present and discuss various aspects of the work presented here. Feedback and support from these events has been particularly welcome. Additionally, I also received helpful comments from audiences at the 2016 San Francisco AAG and Political Geography pre-conference, the 2016 Adelaide IAG conference, and the 2017 Boston AAG and Political Geography pre-conference.

Of course, the act of writing a book extends beyond professional practice, and I have received unwavering support from my family: Jody Wetzel, Rhen, and Smuckers; Elaine and Jim Wetzel; and Larry and Pati Grove have been constant sources of support and inspiration.

Thanks are also due to Andrew Mould, Egle Zigate, and the Routledge staff for extending the invitation to prepare the text and remaining patient while I worked on the manuscript while juggling an overseas relocation. Their editorial guidance has been most helpful, as has the careful eye of Mary Dalton during copy-editing.

Finally, I would also like to acknowledge the following for granting permissions to reproduce material in this book: Dover Publications for

Figure 3.2, reproduced from S. Barr (1964), *Experiments in Topology*; Springer Publications for Figure 3.3, reproduced from C.S. Holling's (2001) *Ecosystems* article, Brian Walker for Figure 3.1, and Ahjond Garmestani for Figure 3.4.

Funding for original research summarized in chapter 6 was provided by a grant from the Mershon Center for International Security Studies at the Ohio State University.

Figures

3.1	Stability landscape representing two basins of attraction	75
3.2	Topological equivalence of a donut and a coffee mug: stretching and de/reforming can turn a donut into a coffee mug (and back). Both represent a solid volume with a single hole	76
3.3	The adaptive cycle, depicting the four phases of ecodevelopment	81
3.4	Panarchy: hierarchically organized, nested adaptive cycles	84
5.1	Olivetti Studio 42 typewriter, Alexander Schawinsky (ca. 1936)	164
7.1	The world of resilience	239

1

RESILIENCE AND GEOGRAPHIC THOUGHT

THE PROMISE OF RESILIENCE

My shoes were not resilient. Although we had only been standing in the rain for fifteen minutes, the mid-afternoon pop-up shower had already soaked them through. I had anticipated this event, dutifully spraying my footwear with a waterproof treatment and packing an umbrella, but to no avail. The deluge overwhelmed my careful planning and transformed my comfortable shoes into waterlogged blister-generating machines. They had lost their form, function and identity in the face of an external shock. Definitely not resilient. But a short distance away from where we stood as I contemplated my situation, the pumps continued humming as cars splashed past. I was standing with a group of scientists and city planning officials in the middle of a Miami Beach road next to one of several new pumping stations the city government had recently begun installing in flood-prone locations. The officials were graciously giving us a walking tour of the city's resilience initiatives. As the rain continued to fall, a city engineer described how the pump collected storm water run-off and pumped it back into nearby Biscayne Bay. After a short slog, we stood in front of a construction zone where crews were raising low-lying roads to drier elevation three feet higher. Between the pumps and road-raising efforts across the low-lying city, Miami Beach's resilience efforts are ambitious, creative and costly. The price tag for the pumps alone topped US$400 million, which the city funded through a monthly surcharge to all residents' sewer and water bills. "The cost of a cell phone bill," one official often quipped at various meetings (field notes, 18 January 2016; see also Cox and Cox 2016). Miami Beach is fortunate in this regard. Notwithstanding pockets of poverty, the city enjoys a wealthy tax base, skyrocketing real estate values, and a booming tourism industry. In an era of global budgeting austerity, this relative prosperity gives the city financial leeway to experiment with different resilience-building efforts. Indeed, many commentators suggest that in the last instance, the pumps will be a futile gesture. The city simply cannot barricade itself behind higher seawalls and pump out the water as the ocean rises, because it sits on top of porous limestone that allows water to seep in from the ground up. If sea levels follow experts' predictions and continue to rise over the next century – and there is no reason to doubt them – the city, much like my shoes, will become a waterlogged, increasingly uncomfortable place to inhabit.

Our tour guides were grappling with an increasingly common challenge: how to design a more resilient city that can survive and thrive despite looming climate change impacts, even if survival involves transforming into a new and unrecognizable aqua-urban form? In this sense, the pumps cannot be dismissed as a futile gesture. They also embody a new kind of promise, the promise of resilience (Aradau 2014): the possibility that even as sea levels rise, even as the world changes around us, we might be able to adjust to these new realities, survive, and even prosper. In a world seemingly overrun by complex threats such as global environmental change or international terrorism that defy modern security strategies premised on spatial division and temporal prediction, this promise has made resilience a cornerstone of both academic research and professional practice. Over the past twenty years, the rising tide of resilience has seeped out of isolated corners in ecology, psychology, engineering and organizational management, and now inundates research across the disciplines. International development and humanitarianism (Chandler 2014a), disaster studies (Manyena 2006; Cannon and Müller-Mahn 2010), urban security (Coaffee and Rogers 2008), environmental management (Gunderson and Holling 2002), community development (Wilson 2012), economic geography (Dawley *et al.* 2010), urban planning (Davoudi 2012) and climate change studies (Eakin and Leurs 2006; Nelson *et al.* 2007), to name but a few, have all seen notable growth in work on and about resilience. Professional fields are no different. Resilience guides emergency planning in the US Department of Homeland Security – whether this involves technical fields of critical infrastructure protection (DHS 2013) or the capacity-building "Whole Community" strategy (FEMA 2011). It shapes decision-making and planning in international disaster management and development frameworks (UN 2005, 2015; DFID 2011), and has thus been exported to emergency management planning agencies around the globe (Grove 2013a). In the UK, it has become the centerpiece of planning for urban security (Coaffee and Rogers 2008; Coaffee 2013), community development (Rogers 2013), regional development (Welsh 2014; Zebrowski 2016), as well as the underlying framework for the turn to Third Way, "soft paternalism" that shapes all manner of governmental practice (Chandler 2013; Chandler and Reid 2016). It lends its name to 100 Resilient Cities, an ambitious philanthropic network sponsored by the Rockefeller Foundation, which seeks nothing less than reforming urban and regional governance within a complex and catastrophe-prone socio-ecological milieu. And of course, it has been the keystone of alternative environmental management strategies stretching back to formative publications on the theory and practice of adaptive management (Holling 1978; Walters 1986; Lee 1993).

This promise of resilience drove us all into the Miami Beach rain. We were conducting background research on integrative and holistic approaches to urban resilience. But we were far from the only group active in the city. A few months earlier, the mayor announced a new partnership with Harvard University's Graduate School of Design, which would give the city a chance to explore innovative design-based resilience initiatives. The city had also recently hired a Chief Resilience Officer and Assistant Chief Resilience Officer, and contracted with design and engineering consultants AECOM to begin drafting a city resilience plan. These moves anticipated the selection of "Greater Miami and the Beaches" for the Rockefeller Foundation's 100 Resilient Cities program. GMTB, as it is now called, is a coalition of Miami Beach, the city of Miami, and Miami-Dade County governments. Their joint selection provides financial support for Chief Resilience Officer positions, facilitates the development of a region-wide resilience plan with guidance from the Foundation and their consultants (including AECOM), and also provides the governments with discounted or free consultancy work from a "menu" of national and international agencies that work on all manner of urban resilience-related issues. In Miami Beach, the demand for resilience is practically unlimited.

There is no such thing as too much resilience, and city officials have a remarkable number of paths – ours included – to explore how they might realize the promise of resilience in practice.

This book will examine the history and implications of this explosive interest in resilience on academic thought and practice in general, and within the discipline of geography in particular. I depart from the assumption that resilience cannot be dismissed as simply the latest academic and policy fad, a flash in the pan that will soon run its course and fall out of favor. Nor can it be written off as mere ideological cover for continued neoliberalization of social and ecological relations, a fashionable charge I will dissect below. Instead, I take seriously the challenge resilience poses to conventional *ways of thinking* about human–environment relations (and socio-spatial relations more generally) and *practicing* social and environmental management. As I will demonstrate throughout this book, resilience is slowly transforming thought and practice in ways that often fly under the radar of conventional forms of analysis and reflection, both critical and applied. My goal is to map these transformations and unpack the opportunities for thinking social and ecological change and uncertainty they introduce – but also their limitations.

I am thus less concerned with what resilience *is* than what resilience *does*. Any attempt to give resilience a clear and concise meaning is problematic, as critics and proponents alike acknowledge (Brand and Jax 2007; Anderson 2015; Dunn-Cavelty *et al.* 2015; Simon and Randalls 2016). Instead, I am concerned with the way resilience thinking transforms the relationship between truth and control. *Truth*: the kinds of things we can know (and not know) about the world around us. *Control*: strategic interventions in the world designed to shape possible outcomes. Since the dawn of European modernity, the relation between truth and control – in the Western world, at least – has been defined as what the French historian Michel Foucault refers to as *a will to truth*: the notion that truth should be objective and lead to reliable prediction, unsullied by subjective values and tastes; predictive knowledge will enable humans to bend nature to their will. But since the early twentieth century, the growing assertion that we live in a complex world beyond total knowledge and control has begun reconfiguring this relation into what I call here *a will to design*. If a will to truth is premised on objective knowledge, a will to design recognizes that knowledge will necessarily be limited or "bounded," to play on behavioral scientist Herbert Simon's (1955) Nobel Prize-winning phrase. For Simon and other complexity theorists, total knowledge of a complex world exceeds human cognitive capacities. Prediction is impossible in such conditions, because complexity implies that change will be non-linear and often unforeseeable. But this does not mean control is impossible. Instead, intervention requires constant monitoring and reflection that will allow humans to adjust (or more precisely, adapt) their interventions to the unpredictable effects that result. A will to design signals the effort or striving to intervene in *and* adapt to a complex world from a position of necessarily partial knowledge.

What does resilience *do* then? I will argue that resilience reconfigures thought and practice on human–environmental relations around a will to design. As I will demonstrate below, Herbert Simon and other scholars of complexity and human artifice shaped the thought of influential resilience scholars, especially C.S. Holling (1978; 2001), Elinor Ostrom (1990; Kiser and Ostrom 1982) and Kai Lee (1993). Although contemporary research has largely written out these influences, excavating their formative role draws out a will to design's continued impact on contemporary developments and debates on resilience. To the extent that the need to become resilient is reshaping thought and action in any number of fields of study and professional practice, then a will to design is the mold around which these activities are re-formed.

However, the ties that bind resilience theory and a will to design are buried deep beneath decades of work on resilience in fields such as ecology and psychology. The proliferation of interest in resilience across the disciplines has only added more dirt to the pile. Excavating these connections and exploring their continued influence requires us to sift through these layers. This means that as we dig through topics and debates that preoccupy current research on resilience, we also must remain somewhat outside or external to them. In other words, we need to *aestheticize* resilience (Groys 2016), or approach resilience thinking as an object of study itself, whose modes of truth-telling about the world and prescriptions for how to live within and manage this world need to be unpacked and analyzed, rather than taken for granted as the starting-point of analysis. The next section considers these points.

AESTHETICIZING RESILIENCE

Resilience is a notoriously slippery concept to pin down. It shifts meaning, function, and efficacy as it flows from one context to the next. Sociologist and disaster studies specialist Kathleen Tierney (2014) recently identified over forty distinct uses of the term in academic and professional literature. While we can trace the concept's roots to engineering, psychology, organizational management and disaster studies, ecologists developed what is arguably the most influential understanding of the term (Welsh 2014; Brown and Westaway 2011). Since C.S. Holling's (1973) path-breaking work on multiple-equilibrium ecosystems, ecologists typically understand resilience as a social or ecological system's capacity to adapt to exogenous disturbances and reorganize itself in ways that maintain existing function and identity (Walker and Salt 2012). This seemingly paradoxical condition of staying the same while changing is one of the reasons resilience has attracted so much attention from geographers and researchers in related disciplines: in a world of complex interconnections and unpredictable, emergent phenomena humans can neither control nor prevent, resilience asserts that change, disruption, and vulnerability are potentially *beneficial* conditions (Walker *et al.* 2004). Resilience focuses on identifying and improving conditions that enable systemic adaptations following disturbances, whether adaptation is a matter of 'bouncing back' to a previous state or 'bouncing forward' to a new and potentially improved configuration.

This understanding of persisting within and through change has made resilience attractive to policymakers and practitioners grappling with different expressions of complex, interconnected social and ecological phenomena. However, this popularity has also sparked a backlash from critical scholars. For example, some disaster studies researchers have argued that resilience brings little new to the table; it simply puts a new name to processes and issues their field has studied for decades (Manyena 2006). In the process, its "scientistic" slant on vulnerability and adaptation detracts attention and resources from underlying socio-economic inequalities that cause vulnerability in the first place (Cannon and Müller-Mahn 2010; see also Gaillard 2010; Pelling 2010). Outside disaster studies, other critical scholars have drawn attention to the depoliticizing ideological and biopolitical effects of resilience-building initiatives (Chandler 2014a; MacKinnon and Derickson 2012; Grove 2013a, 2013b, 2014a; Welsh 2014; Cretney 2014). For these critics, resilience initiatives facilitate the spread and consolidation of neoliberal socio-ecological relations – that is, they normalize vulnerability, seek to make individuals responsible for managing the effects of social and ecological insecurities and disruptions, and thus provide discursive justification for further cuts to social welfare provisioning (Walker and Cooper 2011; Duffield 2011; Evans and Reid 2014). While some researchers suggest resilience might provide a foothold for subversive action (Grove 2013b; Nelson 2014), others argue we should abandon resilience in favor of other concepts with less ideological baggage (MacKinnon and Derickson 2012; Neocleous 2013).

Beneath its waters, the rising tide of resilience generates turbulent debate. Is it a beneficial concept that can help scholars address intractable problems of social and ecological complexity? Does it simply provide neoliberal reforms with ideological justification? And how might we study and make sense of a concept that has clearly become central to research and practice in a number of fields, but that nonetheless is characterized by a *lack* of conceptual clarity and coherence?

Ecologists Fridolin Brand and Kurt Jax's 2007 *Ecology and Society* article, tellingly entitled, "Focusing on the meaning(s) of resilience: resilience as a descriptive concept and boundary object," provides one provocative solution. They begin by reviewing ten distinct ways influential studies have used "resilience," which they then classify into three categories. First, some studies in ecology utilize resilience as a descriptive concept that represents an ecosystem's capacity to reconfigure itself following a disturbance in ways that maintain identity and function. Second, studies in ecology and cognate fields such as environmental studies and geography use resilience as a normative concept that conveys implicit or explicit judgments about how complex social and ecological relations *should* be organized. Third, other publications utilize it as a hybrid concept that blends elements of both descriptive and normative uses. In their diagnosis, this incoherence presents a grave threat to further research: "both conceptual clarity and practical relevance [of resilience] are critically in danger" from the concept's growing popularity in the ecological, social, and biological sciences (Brand and Jax 2007: unpag.). For them, only the descriptive usage found in ecological studies offers a "quantitative and measurable concept that can be used for achieving progress in ecological science" (ibid.). In contrast, other uses fail to prevent normative judgments from coloring objective scientific analysis, and dilute the original meaning of the term in ways that enable ambiguous and divergent uses. They propose a so-called "division of labor in a scientific sense" (ibid.) in response. On one hand, ecologists – and ecologists alone – should be able to lay claim to resilience as a descriptive concept, since a purified, objective and descriptive understanding offers a "clear, well defined, and specified concept that provides the basis for operationalization and application within ecological science" (ibid.). But this does not mean other disciplines cannot contribute to research on complex social and ecological systems. They also suggest that normative and hybrid uses "foster communication across disciplines and between science and practice," a trait they see as quite desirable (ibid.). Thus, to promote integrative, cross-boundary research, they suggest that researchers outside of ecology should utilize resilience as a "boundary object" that enables interdisciplinary analysis of complex social and ecological systemic dynamics.

Brand and Jax's formulation of resilience as a boundary object certainly helps us conceptualize how resilience has emerged from provincial corners of ecology and psychology (among other fields) to become one of the most influential concepts in contemporary social and environmental research (Zimmerer 2015). However, for my interests here, their efforts to promote resilience as a means of facilitating cross-boundary research point us towards a distinct *style of thought* about social and environmental change, a certain way of posing problems and devising solutions to these problems that I am here calling a will to design.

RESILIENCE AND A WILL TO DESIGN

With the concept of *will to design*, I am signaling an ethos of sorts, a drive or desire to *synthesize* otherwise distinct forms of knowledge in order to develop collaborative and cross-boundary solutions to contextually specific problems of complexity. My use of this term draws heavily on work in design studies. At first glance, this may appear to be an odd pairing. For the most part, design is an alien concept in geography. The term does not have an entry in the encyclopedic *Dictionary of Human Geography* (Gregory *et al.* 2009). And yet, as I hope to show throughout this book, the path towards understanding how resilience is shaping contemporary possibilities for thought on social and ecological change takes us through the cybernetic roots of design theory.

The influential design studies scholar Victor Margolin defines design in the following terms:

> [d]esign is the *conception and planning of the artificial*, that broad domain of human made products which includes material objects, visual and verbal communications, organized activities and services, and complex systems and environments for living, working, playing, and learning.
>
> (Buchanan and Margolin 1990; quoted in Margolin 1995: 13, emphasis added)

Margolin's definition is particularly useful because it encourages us to look beyond conventional understandings of design as something concerned with surface appearance, aesthetics and subjective taste – the formal qualities of objects, in other words – to distinct *processes* involved in "conceiving" and "planning" human artifice. Since the mid-twentieth century, design studies scholars have gone to great lengths differentiating design as a style of thought and practice distinct from both the sciences (physical and social) and humanities (see Box 1.1). For example, in a well-cited essay in the journal *Design Studies*, Nigel Cross (1982) outlines what he calls "designerly ways of knowing." With this phrase, Cross argues that design is characterized by "a synthesis of knowledge and skills from both the sciences and humanities, in pursuit of practical tasks; it is not simply 'applied science', but the application of scientific and *other organized knowledge* to practical tasks" (Cross 1982: 222, emphasis in original). This is by now a common refrain in design studies: design is not concerned with analysis and the production of truth and certainty; rather, it synthesizes different forms of knowledge in order to solve immediate and practical problems. This solution may take a variety of forms – a material artifact such as a building or an instrument, a semiotic device such as lines of software code or a branding strategy, an artistic production such as a musical composition, or even a course of medical treatment "designed" to cure an illness – but what unites all of these design products is a set of more or less shared information handling procedures that create the solutions in question (Rowe 1987). Indeed, in his field-defining 1996 work, *The Sciences of the Artificial*, Herbert Simon positions design at the root of intentional human action: "everyone designs who devises courses of action aimed at changing existing situations into preferred ones" (Simon 1996: 111).

Box 1.1 A BRIEF HISTORY OF DESIGN STUDIES

Much like resilience, design has attracted increased attention in recent years, particularly among the management professions Top-selling titles profess to reveal how design thinking can improve organizational performance and drive innovation (e.g., Brown 2009). However, this popular literature is epistemologically distinct from the forms of *designerly thinking* developed with the discipline of design studies (Johansson-Sköldberg *et al.* 2013). Nigel Cross, longtime editor of the journal *Design Studies*, describes the discipline's focus in the following terms:

> Our concern in design research is the development, articulation, and communication of design knowledge. Our axiom has to be that there are forms of knowledge peculiar to the awareness and ability of a designer, just as the other intellectual cultures in the sciences and the arts concentrate on the forms of knowledge peculiar to the scientist or the artist
>
> (Cross 1999: 5)

His definition highlights how design theorists differentiate their field of study from other academic pursuits by claiming a distinct form of knowledge as their object of study. If science analyzes the natural world and art examines the human experience, then design explores creativity, imagination and practicality. Design studies thus investigates the creation of artifice, or the production of objects "designed" to meet specific practical needs.

While the modernist design movement was arguably inaugurated with the 1919 opening of the Bauhaus school in Weimar (see Chapter 5), the academic study of the design process only began emerging in the 1950s through the work of avant-garde architectural critics and art historians such as Reyner Banham (Margolin 1995). Prior to this era, modern designers believed that meaning simply inhered in objects. However, Banham and others showed how the meaning of artifacts was constantly negotiated through shifting and indeterminate relations between the user and the object. This meant that there was always a history to mass-produced goods. And if we follow Victor Margolin's (1995: 15) suggestion that "the product is located in a situation and its meaning is created in part by its users," then design studies could examine the creative process of constructing artifacts and imbuing them with meaning.

However, the quantitative wave that subsumed many social science fields in the 1960s also caught up design studies (on the similar spatial-quantitative revolution in geography, see Gregory 1978; Johnston 1979; Cox 2014). Quantitative and cybernetically minded design scholars, drawing inspiration from Herbert Simon's (1955) work on bounded rationality and decision-making in complex choice environments (see Chapter 5), strategically positioned design as the scientific study of universally applicable procedures for creative problem-solving. Design studies increasingly involved attempts to model the creative process as a stepwise, adaptive search for solutions to complex problems. While this enabled scholars to demarcate a unique academic field and endow it with the legitimacy of a scientific endeavor, for a growing chorus of critics, it also bore little resemblance to the actual process of design or the skill sets designers drew on to formulate and address problems. Horst Rittel's work on *wicked problems* (Rittel and Webber 1973) was a key intervention here. Wicked problems are a specific class of problems. They are not simply ill-defined problems that lack specification in terms appropriate for a particular branch of science or engineering to solve. Instead, they are problems that *cannot* be fully determined (see also Coyne 2005). Like environmental pollution, urban redevelopment, sustainability or (more recently) resilience, they often transcend disciplinary boundaries and involve stakeholders with competing and incompatible interests and values. Any proposed solution will thus inevitably generate additional and unforeseen problems, which means the problem lacks a stopping rule for terminating the search for solutions. Similarly, any proposed solution largely hinges on how the designer chooses to frame the problem in the first place, which means multiple, equally valid solutions are possible.

Rittel and Webber's intervention thus paved the way for alternative visions of design based on constructivist or hermeneutic approaches concerned with *how* designers engaged with indeterminacy (e.g., Schön 1983). While a number of distinct approaches have developed since the 1980s (see Cross 1999, 2007 and Johansson-Sköldberg *et al.* 2013 for overviews of design studies' various "schools"), one example comes from Richard Buchanan's influential theory of *placements* (Buchanan 1992, see Chapter 6). Buchanan's concept of placement draws on both Simon and Rittel and Webber to describe any device designers used to re-position a problem, or view it from different perspectives (Buchanan 1992; see Chapter 6). This creative and "play-ful" (Wylant 2010) re-contextualization enables designers to experimentally synthesize different perspectives and develop contingent, contextually specific, pragmatic solutions to otherwise intractable problems.

Because wicked problems are by their nature indeterminate, no single academic discipline can claim to hold a definitive solution. Instead, the designer must pragmatically synthesize distinct and perhaps incompatible forms of knowledge. Design studies, like geography, is thus an inherently interdisciplinary and integrative discipline. But where geography integrates the social and physical sciences to develop detailed *analytical understandings* of space and place and human–environment relations, design *synthesizes* across the sciences (social and physical) and humanities to develop *pragmatic solutions* to wicked problems. For Buchanan,

> Design problems are "indeterminate" and "wicked" because design has no special subject matter of its own apart from what a designer conceives it to be. The subject matter of design is potentially universal in scope, because design thinking may be applied to any area of human experience. But in the process of application, the designer must discover or invent a particular subject out of the problems and issues of specific circumstances.

Barry Wylant (2010: 228) further explains that,

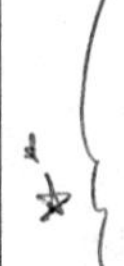

> Design thinking is really the discipline then of cycling through many such contextual exercises to understand how sense *can* be made of something and given this, the designer is then in a position to choose which contexts *should* dominate and the manner in which they should.

Wylant's specification of design as a practice of sense-making provides us with a succinct description of design studies in general. If we position his description in relation to design studies' origins in 1950s art criticism, as above, then we can see how the field has quickly matured into a distinct academic discipline that addresses the integrative and trans-disciplinary process of sense-making that enables designers to give provisional yet determinate form to indeterminate problems. As we will see in Chapters 5–7, the practices and sensibilities designerly thinking mobilizes to engage with indeterminacy implicitly shape resilience scholars' own engagements with social and ecological complexity.

I will return to Simon below, because he played an indispensible role in specifying the distinct style of thought design and resilience animate. For now, with this wide-ranging and process-oriented understanding of design in mind, in this book I approach design as an art of the threshold: *an art of relating an interior to an exterior, in the absence of any transcendental determinants of value*. For example, a fashion designer designs clothing that enables the wearer to present a certain image of her physical appearance to the judgmental gaze of the Other (Groys 2008). A landscape architect designs an urban park to improve quality of life in the context of conflicting state, private and public interests and land use preferences (Marshall 2013). Or in perhaps more familiar terms, community members design property regimes to govern a population's access to scarce common-pool resources (Ostrom 1990). In each case, there is no way of knowing what the "right" and "wrong" design will be – perfect knowledge of subjective tastes and emergent environmental, social, political, and biophysical conditions is prohibitively complex (Simon 1996). Instead, a design's effectiveness will only become apparent as clothing is worn, as plans for a park work their way through various bureaucratic approval processes, as community members utilize common-pool resources, and so on.

The designer thus does not inhabit the infamous "view from nowhere" (Haraway 1988), a position of detached scientific objectivity that provides purportedly objective knowledge, prediction and control. This is a key point. The scientist who attempts to inhabit this view from nowhere expresses a will to truth that seeks to reduce all physical and human phenomena to discrete, quantitative, calculative (algebraic) relations that can be known with certainty, predicted, and thus controlled (see Foucault 1989; Elden 2001). A will to truth has animated modern science's unending drive to discover the universe's hidden laws and use this knowledge – this truth, as it were – to perfect both the natural and social worlds (Gregory 1978; Haraway 1997). But as I indicated above, a *will to design* is distinct from a will to truth. The world the designer inhabits, according to design scholars, is too complex for an individual to know, understand and manage (Schön 1983). There are too many interconnections between individuals, across sectors, between individuals and environments, across environments, within and through organizations, *inter alia*, and too much information passes through these interconnections for any one individual to grasp. Any "truth" we purport to produce will be partial and limited, and any attempt to control the world through this partial and limited knowledge will only lead to failure and hardship (Rowe 1987).

Herbert Simon (1955, 1956) characterizes this limited condition as "bounded rationality." He developed this concept as a sympathetic critique of neoclassical economic theory, which posits a rational individual decision-maker with perfect information – the infamous *homo economicus*, or "economic man" (see Foucault 2008). In contrast, Simon asserts that decision-makers are embedded in complex environments that human rationality cannot fully understand or process. He argues that decisions taken in the absence of total knowledge are not irrational, but are rather rational within contextual limitations. They are pragmatic and shaped by habits and institutions, which provide the decision-maker with "reasonable" expectations for the kinds of results that might arise from one course of action or another. In this light, we might understand Simon's theory of bounded rationality as "neoliberal" – *not* because it involves elements of privatization, individualization, deregulation, and marketization of social and ecological relations (a common trope in critical research on neoliberalism; see Barnett 2005) – but rather because he attempts to update key tenets of classical liberal political economic thought within the context of novel experiences of space and time (Collier 2011). As I will detail below, developments during the early twentieth century in fields such as military strategy, the life sciences and computational sciences gave rise to new understandings of life as inherently interconnected, emergent, and adaptable (DeLanda 1991; Martin 2007; Collier and Lakoff 2008; Dillon and Reid 2009; Sloterdijk 2009; Evans and Reid 2014). These are precisely the conditions of complexity Simon grappled with, and the neoliberal qualities of his contribution lie in devising a theoretical framework that shows how behavior can still be rational, even if this form of rationality is alien to modern liberal notions of the term.

To be sure, Simon's life's work embodied modern scientific conceit: he was dedicated to uncovering laws that govern human decision-making, even if the decision-maker is embedded in a complex world (Crowther-Heyck 2005; Buchanan 1992). But his specifications of bounded rationality, adaptation and hierarchy helped formalize synthetic knowledge production animated around a different kind of will, a will to design, that makes truth function in new ways. For the individual embedded within a complex environment, truth no longer conveys transcendent, invariant laws that govern physical and social phenomena. Instead, truth is an effect of bounded rationality, a necessarily partial understanding of the world that can *contribute to*, but not *determine*, a better understanding of complex systemic dynamics – but only if it can be pragmatically combined with *other, equally valid* truths generated through *other* bounded rationalities. A will to design is precisely this attitude, sensibility, or more precisely, desire to break down boundaries (an inherently Simonian endeavor – see Crowther-Heyck 2005) and engage in collaborative, pragmatic work that constructs solutions to complex problems individual actors are constitutively incapable of grasping on their own.

Simon's work influenced a generation of scientists, economists, psychologists, computer programmers, administrators, and designers seeking to better understand and manipulate how individuals respond to changes in complex environments (Jones *et al.* 2013; Chandler 2016a). His conceptual innovations laid the groundwork for many key concepts and techniques now associated with resilience, such as panarchy and adaptive management. This influence is particularly clear in the work of three scholars who developed many theoretical foundations for contemporary debates on resilience and environmental governance: C.S. Holling, Elinor Ostrom and Kai Lee. Holling often singled out Simon's work on hierarchies for influencing ecologists' understanding of complexity, self-organization and the adaptive cycle (Holling 1978, 1986, 2001). Similarly, Ostrom (2009: 2) credited Simon and her partner Vincent Ostrom as important theoretical influences for her own Nobel Prize-winning research on institutions of collective governance. And Lee (1993) based his understanding of adaptive management on Simon's theory of bounded rationality and satisficing decision-making behavior. Thus, the roots of resilience can, to a surprising and largely under-acknowledged extent, be traced back to earlier scholarly efforts to grapple with humans' fundamental creativity and unpredictability in a complex and indeterminate world lacking transcendental guarantees.

This same will to design animates Brand and Jax's (2007) efforts detailed above to clarify the concept of resilience. Their specification of resilience as a descriptive concept and boundary object, and the resulting "division of labor in a scientific sense" recognizes that social and ecological systemic dynamics are too complex for a single discipline to know and predict. Of course, on one level they maintain optimism that resilience as a descriptive concept can enable ecologists (and ecologists alone!) to develop more precise and accurate knowledge of a specific object of study: complex ecosystems. But their specification of resilience as a boundary object de-centers this ecological will to truth. It does not attempt to discredit competing understandings of resilience and assert one true definition. Instead, it recognizes that different forms of expertise each bring something to the table. The "bounded rationality" of disciplinary knowledge is too limited to understand and control complex social and ecological systems, but synthesizing different knowledges can elucidate novel dimensions of this complexity. In short, Brand and Jax transform a problem of epistemology – too many definitions of resilience that threaten to dilute its utility – into a problem of design: how might these different bodies of knowledge on "resilience" work together in productive ways? Reworded slightly, we could pose Brand and Jax's problem in design thinking terms: how can resilience productively repurpose and synthesize different bodies of research with the goal of better understanding and managing complex social and ecological systems? This is, of course, a problem without a specific, optimal answer. Instead, it is a design opportunity – a dislocation between an inside (knowledge) and an outside (complex systems) that can be addressed through specific design interventions (cross-disciplinary work mediated through resilience as a boundary object). It does not seek to uncover objective truths that enable predictive knowledge and more perfect control of social and environmental phenomena. It does not express a will to truth. Instead, it expresses a will to design, a desire to pragmatically and collaboratively engage with complex phenomena from a position of necessarily partial and limited knowledge.

In many ways, a will to design is an important departure from positivist social and environmental research that critically minded geographers have traditionally taken as the foil of their critique. Collaborative, integrative and interdisciplinary research is precisely the kind of scientific practice that sustainability scientist Thad Miller and colleagues call "solutions-oriented" research (Miller *et al.* 2014). A pragmatic, solutions-oriented ethos has also been at the heart of growing interest in so-called "experimental" approaches to adaptation, urban resilience and socio-technical-ecological transitions, which emphasize the need to test out, in living labs of cities, innovative solutions to problems of complexity (see Evans 2011; Turnpenny and O'Riordan 2007). This live-action experimental and solutions-oriented research is a form of science that is useful to policymakers, to be sure (and the "utility" of scientific knowledge is one pillar of positivist scientific practice; see Gregory 1978), but its claims to utility are not grounded in scientists' professed ability to develop objective, predictive knowledge. Indeed, despite critics' assertions that resilience approaches are overly "scientistic" (Cannon and Müller-Mahn 2010), ecological understandings of resilience also recognize *the limits* of scientific knowledge: echoing Simon's understanding of complexity and bounded rationality, many ecologists themselves recognize that the inherent complexity of social and ecological systems means that no one discipline can provide perfect knowledge on systemic dynamics (e.g., Holling 1986; Lee 1999; Folke *et al.* 2002; Walker and Salt 2012). Faced with such complexity, the usefulness of science can no longer be guaranteed through its claims to a privileged understanding of The Truth. Instead, usefulness emerges out of the *process* of collaborating on complex problems that have no straightforward, predictable solutions. It is contingent on pragmatically identifying ways that different (and limited) forms of scientific knowledge can advance collaborative efforts to (re)define specific problems that need to be addressed, develop different kinds of interventions, monitor the effects of these interventions to ensure they are having desirable impacts on social and ecological dynamics, and adjust these interventions as needed.

The turn to solutions-oriented, experimental science of the sort espoused by Miller and colleagues, Brand and Jax and others thus mobilizes a will to design that attempts to contingently assemble diverse forms of knowledge and interests in ways that can address specific problems of complexity. This is an exciting development. A will to design encourages scientists and practitioners to remove themselves from disciplinary and professional silos and recognize the value of perspectives that differ from their own. It creates new possibilities to fold in different forms of knowledge and styles of reasoning that have often been left out of policymaking processes and the production of scientific knowledge. But this is not without problems. Most importantly, as I detail in Chapter 6, an emphasis on synthetic rather than analytic reasoning transforms the meaning of critique. A solutions-oriented intervention might position itself as critical because it is attempting to address problems of socio-economic inequality through innovative, collaborative solutions. But synthesizing diverse forms of knowledge to develop a pragmatic solution is a different form of critique than the analytical critiques critical geographers have traditionally offered. If analytical critique attempts to develop new understandings of how the world works, with an eye towards dislodging assumed truths and making it possible to transgress present conditions and reinvent the world, the synthetic critique of solutions-oriented research demands that knowledge work with the world to improve how things are (Boltanski 2011; Chandler 2014a). A will to design does not promise revolution, but rather pragmatic improvement (Latour 2008), often visualized as a managed and regulated transition to new forms of socio-ecological order (Kemp and Rotmans 2005). In turn, a pragmatic imperative to produce innovative solutions and guide transitions introduces new judgments on the value of different forms of knowledge: a will to design implicitly demands that *a body of knowledge must offer some kind of pragmatic utility* to those who are not a part of its disciplinary debates. In other words, a different form of knowledge is valued to the extent that it is translatable and amenable to synthesis with other forms of knowledge – and translation always carries with it the implicit threat of eroding difference (see Povinelli 2006, 2011a).

But I am getting ahead of myself. For now, what matters is that Brand and Jax's arguments provide a window onto an unconventional yet provocative understanding of resilience theory. Resilience introduces a distinct style of thought that subtly transforms the meaning, significance and value of both critical and applied research. This book seeks to explore this style of thought and the challenges and opportunities it presents for geographic thought specifically, and critical theory more generally. The next section details the main arguments I will develop along the way.

GENEALOGIES OF RESILIENCE

To analyze the relation between resilience and geographic thought, I adopt a *genealogical* style of analysis. While Chapter 2 will detail what genealogy entails, for now, we can succinctly characterize it as a mode of thought that questions how conditions we commonly take for granted came about (Foucault 1977a, 2003). In the case of resilience, such conditions include the apparent complexity of human–environment relations, and the need for creative, innovative, synthetic solutions to complexity. A genealogy focuses on developments both within and beyond the boundaries of disciplinary fields of expertise to examine how truth claims have been shaped by wider social, cultural, political, economic, and environmental processes. It is thus a particularly appropriate mode of analysis for unpacking the emergence of resilience, because it enables us to position resilience within wider efforts to understand and regulate novel experiences of spatial interconnection and temporal emergence. As I indicated above, we can trace the roots of resilience to the emergence of a modern design aesthetic and debates over complexity and cybernetics in fields such as military planning, computer science, public administration, political economy and behavioral sciences. A genealogical analysis charts the continuities, transformations, reconfigurations, translations and dislocations across these otherwise diverse fields to demonstrate how the forms of knowledge, styles of thought, techniques for monitoring and calculating dynamic phenomena, and practices of regulating these phenomena that developed out of this work shaped the possibilities for contemporary research on social and ecological complexity.

On the basis of a genealogical analysis of resilience, I will develop two interrelated arguments. *First*, I will argue that resilience recalibrates key pillars of geographic thought – specifically, the study of human–environment relations and socio-spatial relations – around a designerly aesthetics. That is, resilience transforms the possibilities for sensing and perceiving the world. This recalibration has ontological, epistemological, and ethical effects. On the level of ontology (that is, our understanding of what exists), resilience thinking asserts that human–environment relations and spatial relations are complex, interconnected, and emergent. It also implicitly assumes that what exists can be reduced to rational abstractions (see Baudrillard 1981) and functionally synthesized with one another to produce systemic effects. On the level of epistemology (our understanding of what knowledge is and how knowledge can be generated), resilience thinking asserts that any claim to truth is partial and limited, and thus the production of knowledge about complex phenomena should be a collaborative process. On the level of ethics (our understanding of how we should live within the world and relate to other human and non-human entities), resilience thinking inculcates a will to design, a desire to synthesize different forms of knowledge and create pragmatic solutions to complexity. Together, these three levels, or more appropriately, vectors, comprise what we might think of as the *diagrammatic* characteristics of resilience. A diagram is as an abstract, ideal matrix that structures dynamic and differential force relations. With its basis in Deleuzian philosophy (Deleuze 1988), the concept of diagram assumes that human – environment relations are indivisible: it expresses an ontology of immanence, which posits "what exists" as relations that give rise to bodies (individual, collective, human and non-human) with certain capacities and desires. Diagrams attempt to structure these relations, and thus attempt to shape capacities and desires. In essence, they attempt to turn an immanent, immeasurable field of relational potentiality into a determinate series of possibilities (Deleuze 1995a).

While this definition of diagram might seem overly arcane, it signals something basic: as a diagram, resilience is a way of intervening within debates over how to understand and manage interconnection and emergence. Through concepts such as panarchy (Holling 2001; Gunderson and Holling 2002), resilience structures immanent human–environment relations as nested sets of complex social and ecological systems whose cross-scalar interactions shape responses to external stressors. This specification opens up a field of possibilities for understanding and managing processes of social and ecological change. For example, change can be understood as maladaptation that increases the possibility of systemic collapse, or proper adaptations that enable a system to develop and improve its performance, however this might be measured. Techniques such as adaptive management and adaptive governance enable researchers and stakeholders to manage this process of change and increase the possibility of desirable outcomes – that is, proper adaptation.

As a diagram, resilience has no essential meaning. Diagrams gain their meaning and significance only as they are mobilized in specific contexts to address specific problems of thought. This means that resilience can manifest itself in different ways across a variety of settings outside narrow, discipline-focused research in fields such as ecology or psychology. This mobility and pliability allows researchers, policymakers, practitioners, and lay persons to make sense of their worlds in new ways through the concept of resilience: to envision themselves as part of complex systems they can neither fully know nor completely control; to pose uncertainty and insecurity as problems of how to live with and manage complexity; and to experimentally design certain ways of acting together and intervening in complexity in order to bring about desirable outcomes. This plasticity does not mean that resilience lacks meaning; instead, what resilience is is a matter of *how resilience intervenes* in contextually specific debates over social and ecological change, and the precise *effects* that result from this intervention (Anderson 2015). As a diagram, resilience is more than simply a descriptive concept that currently lacks precise meaning, or a boundary object that facilitates collaborative solutions (Brand and Jax 2007). It is both of these, to be sure, but it is *also* other things at one and the same time. Resilience is a descriptive concept, *and* a boundary object, *and* an ideology, *and* a discourse, *and* a metaphor, *and* a governmental rationality, *and* a regime of truth, *and* an idiom, *and*... *and*... The list could stretch on. Resilience is all of these and more precisely because it is an abstract relation between humans, environments, and technologies that has been mobilized in specific contexts and debates to produce effects that we then attribute to a descriptive concept, or a boundary object, or an ideology, or a discourse, or a metaphor, or so forth.

However, this abstractness does not mean that we cannot give resilience positive content. Instead, as I hope to show, various mobilizations of resilience all express a designerly style of thought and practice that attempts to engage with the world from a position of necessarily limited knowledge and control. The genealogical analysis I offer here will trace how this will to design took its contemporary form in mid-twentieth-century critiques of centralized planning, how it helped shape ecological approaches to resilience in the 1970s, and how it has come to inflect key geographic traditions such as human–environment relations and spatiality.

Second, along these lines, I will argue that a will to design presents challenges and opportunities for both applied and critical geographic thought. On one level, a will to design creates new opportunities for applied and critical scholars to contribute geographic insights to cross-disciplinary debates on social and environmental change – topics that have long interested geographers. But there is a heightened risk that a will to design facilitates other disciplines laying claim to areas that fall within the traditions of geographic thought. This disciplinary creep is particularly evident in early claims from ecologists and systems scientists that complexity science holds the key to understanding human–environment relations (Kaufmann 1995, 2000; Holling 2001, 2004; Holling and Sanderson 1996). Even though a will to design preaches the gospel of interdisciplinarity, it still functions in an institutional world of distinct academic disciplines, and the *effects* of this work can spark new disciplinary fault lines – for example, by impacting research funding and publishing landscapes. The rush to engage in collaborative research can inadvertently cede the epistemological high ground to other disciplines, possibly (although not necessarily) undermining the discipline's standing. In such a condition, geographers need to become more strategic and reflexive in how they engage with resilience in order to ensure that cross-disciplinary work enhances rather than erodes the value of geographic knowledge.

On another level, the emergence of a will to design presents a new challenge for critical geographic thought. Since its inception in the early 1970s, critical geographic work has been calibrated around a more or less implicit critique of a will to truth. Critical geographers have dedicated considerable time and energy to demonstrating the *partiality* of totalizing knowledge claims and their often unacknowledged political biases and effects – whether these are knowledge claims in areas such as development (Smith 1984; Blaikie and Brookfield 1987; Peet and Watts 1993; Leach and Mearns 1996; Rocheleau *et al.* 1996), geopolitics (Dalby 1991, 2002; O Tuathail 1996; Hyndman 2001), urban studies (Harvey 1973, 1985), hazards (O'Keefe *et al.* 1976; Watts 1983; Hewitt 1983), economic development (Scott 1988; Storper 1997), and numerous other sub-fields. This style of critique has informed a number of recent engagements with resilience that have demonstrated its close ties to neoliberal development and its insidious depoliticizing effects (MacKinnon and Derickson 2012; Cretney 2014; Welsh 2014; Grove 2013a, 2013b). These studies all show how resilience consolidates rather than challenges the political ecological status quo. But to an extent, these critiques overlook the synthetic style of reasoning that underpins resilience interventions. Resilience thinking's underlying will to design does not displace a will to truth, but rather mobilizes truth towards a new end: neither prediction nor total control, but rather designing solutions that are necessarily partial and limited, and that will have to be revisited and revised as new problems arise. The modesty and pragmatism of resilience poses a distinct challenge for geographic work: what are the possibilities for critical geographic research when the outcome of this research – the demonstration of difference and the partiality of knowledge claims – is precisely what a will to design values and seeks to incorporate into pragmatic, solutions-oriented interventions? In other words, what happens when critique becomes the motor rather than the saboteur of governmental practice?

A number of authors in security studies and international relations have developed theoretically sophisticated critiques of resilience's depoliticizing effects along these lines (Walker and Cooper 2011; Aradau and van Munster 2011; Duffield 2011; Chandler 2014a; Evans and Reid 2014). Although each author provides a distinct analysis of resilience, they share a common concern that resilience thinking increasingly inhibits individual and collective capacities to act politically – that is, to imagine and bring about possibilities for different forms of life not organized around institutions such as private property, the market, and the territorially based nation-state. And this is certainly one of design's problems signaled above: a will to design seeks to improve the present rather than destabilize and reinvent the world. But I do not think the (anti-)politics of resilience are as straightforward as this. If we approach resilience simply as a discourse, or a unified and coherent governmental rationality, or an ideology, as has been the tendency in these critical analyses, then the picture does appear bleak. However, my hope is that focusing on the diagrammatic qualities of resilience and its underlying will to design will enable us to develop a more complex account of resilience's political possibilities.

To foreground both the challenges and opportunities resilience poses for geographic thought, my analysis throughout this book seeks to draw out, detail and destabilize emerging styles of reasoning that are institutionalizing a will to design within contemporary research imperatives. At stake here is the possibility for work in geography and cognate fields such as anthropology, sociology, international relations, and security studies to engage with interconnection and change *beyond* the solutions-oriented, collaborative ethos that increasingly permeates contemporary modes of social and ecological governance.

This text thus offers an analysis of resilience that takes the present state of affairs in geographic thought as a moment for creative and transgressive reflection. Over the past decade, resilience has clearly emerged as *the* definitive concept for understanding novel experiences of space and time in terms of complex interconnection and emergence. The understandings it enables also open social and environmental relations to novel forms of control. But the critical ethos of my analysis maintains that this is not the only possible response to complexity. As Chapter 7 discusses, geographers and researchers in allied fields have developed a range of concepts and theoretical and methodological frameworks for understanding and engaging with spatial complexity and temporal uncertainty that are not oriented around the solutions-focused ethos of a will to design. I will suggest that the problem of resilience hinges on how we deploy topological thought. In brief, topology signals a mode of thought that is attuned to *process* of change, becoming, and impermanence, rather than stability and permanence. As Chapter 3 details, it is one of the defining characteristics of ecological approaches to resilience. But geographers and other critical scholars have long mobilized topological sensibilities to rethink socio-spatial relations (Massey 2005; Amin 2002; Amin and Thrift 2002), human-environment relations (Whatmore 2002), and the operation of power relations more broadly (Collier 2009; Allen 2011). This suggests that topological thinking is politically pliable. As Lauren Martin and Anna Secor's (2013) recent review suggests, topology can both reinforce and transgress structural determinants of meaning and value. Resilience thinking thus opens the problem of topological thought: on the one hand, it mobilizes topological sensibilities to render all instances of social and ecological change as expressions of an underlying structure of hierarchically organized complexity. It tends to read change as either adaptation or maladaptation, and justifies managerial interventions into human–environment affairs in order to ensure "proper" forms of adaptation. In this case, topological thought captures and constrains the destabilizing implications of complexity (Pelling 2010).

But on the other hand, a variety of work in geography, anthropology, international relations, security studies, and cultural studies – to name a few – mobilizes topological thought to destabilize the present. Some of the key movements here include the ontological turn in anthropology (Viveiros de Castro 2014; Povinelli 2011a, 2016), geographers' related interests in ontological politics (Whatmore 2013; Lorimer 2015; Simon and Randalls 2016), assemblage theory (DeLanda 2006; Anderson *et al.* 2012); Guattarian ethico-aesthetic practice (Kanngieser 2013; Evans and Reid 2014; Gerlach and Jellis 2015; Grove and Pugh 2015; Pugh and Grove 2017), work on the radical potential of hacking (Wark 2004; Galloway 2007; Chandler 2016b), and even radical design studies (Manzini 2015; Escobar 2018). Importantly, these approaches all share in common a concern with *transgression*: with identifying, subverting and destabilizing the institutional, conceptual, and theoretical formations that delimit how we are able to think and act. Each attempts, in their own way, to demonstrate the existence of difference *within* the present – that is, the real possibility for things to be other than they are. In many ways, transgression runs at a cross-current to design: if a will to design strives to synthesize different and potentially opposing elements in order to develop solutions to a specific problem, a will to transgress demonstrates the partiality of these solutions. It mobilizes a will to *other* truths, an ethical commitment to the continued persistence of multiplicity and difference within the present – and the possibility that this difference will be *resistant* to synthesis and translation. If design is solutions-oriented, transgression dis-orients solutions. But while this dis-orienting effect is commonly taken in negative terms – as a deconstructive move that inhibits action – it can also be a source of constructive provocation. Transgression shows that any solution will necessarily have "negative" effects – negative not in a moral sense of bad or evil, but rather in the sense that bringing one particular vision of the future into being (that is, into the present) will necessarily limit or even negate the possibility for *other* visions of the future (Povinelli 2011a).

In demonstrating that things could always be other than they are, transgression shows us that there is still always work to be done.

The spirit of this book is in many ways a transgressive one: in drawing attention to *other* ways geographers and researchers in allied fields engage with interconnection and emergence, my hope is to open new possibilities for engaging with resilience and its underlying will to design in creative, playful, inventive and potentially subversive ways. This book strives to chart the terrain of thought on resilience in a way that enables geographers to recognize and transgress the will to design that increasingly shapes how scholars, applied and critical alike, can think about and address social and ecological problems.

THE STRUCTURE OF THE BOOK

The relationship between resilience and geographic thought can (and has) been explored in a variety of ways. Inevitably, this diversity means that the tack I take here will differ from the one that others might take. To take perhaps the most obvious difference, I do *not* intend this book to provide an exhaustive review of the different ways geographers have engaged with resilience. Several excellent reviews of resilience and geography from a variety of perspectives have been published in recent years (see Pelling 2010; Brown and Westaway 2011; MacKinnon and Derickson 2012; Cote and Nightingale 2012; O'Brien 2011; Welsh 2014; Cretney 2014; Weichselgartner and Kelman 2014; Walsh-Dilley and Wolford 2015; Simon and Randalls 2016; Brown 2016). These reviews have provided a handle on the veritable explosion of publications on resilience, and have given some sense and direction to debates over an unwieldy concept. At this point, any attempt to review all the ways geographers have utilized resilience would only duplicate these efforts. Instead, my interest here is to build on this work by identifying and unpacking the specific style of thought associated with resilience that shapes how researchers pose problems that are worthy of study, devise ways and methods to study these problems, and utilize the knowledge these studies produce.

Chapter 2 sets the stage through an overview of current debates on resilience in geography and cognate fields. Rather than attempting to pin down a specific definition of resilience, my interest here is to demonstrate the wide range of definitions in play, and the variety of social and environmental management programs that mobilize these definitions. This uneven terrain suggests that resilience is an *essentially contested concept*: resilience does not merely have a variety of definitions; more importantly, these definitions all convey competing normative claims. This means that any definition of resilience is at once ethical and political: it carries implicit prescriptions for how people and things can and should exist in a complex world of interconnection and emergence. To understand what resilience does thus requires engaging with it in a way that foregrounds rather than sweeps aside these ethical and political considerations. As I show, genealogical work on resilience offers a unique slant on this problem. Its emphasis on contingency and conflict directs analytical attention to the way that specific understandings of resilience emerge out of historically specific struggles over understanding and regulating human–environment relations.

After detailing some elements of genealogical thought and their relevance for resilience, Chapters 3 and 4 offer a preliminary first-cut genealogy of resilience. Chapter 3 examines resilience thinking in ecology as a distinct mode of topological thought attuned to change and becoming rather than permanence and stability. Set within the context of debates over environmentalism and economic development in the 1970s and 1980s, this mode of thought was far outside mainstream approaches to environmental science and management. Chapter 4 unpacks how resilience scholars engaged with this conventional work through two lines of critique. First, they identified and attacked what they saw as the "pathology of command and control" (Holling and Meffe 1996), or the tendency for centralized emergency management strategies, organized around the principle of optimizing resource levels, in practice often resulted in systemic collapse. Second, they raised the question of "institutional fit" (Folke *et al.* 2007), which identified a mismatch between, on the one hand, complex, non-linear and dynamic systems, and on the other, social institutions organized around principles of stability and predictable, linear change. On the basis of these twin critiques, resilience scholars argued for new forms of environmental management founded on the topological principles of resilience theory.

The emergence of resilience thus signals a critical break from established forms of social and environmental management: it seeks nothing less than reconfiguring thought and practice on human–environment relations around what Holling (1973: 15) calls a "very different view of the world," a topological sensibility attuned to systemic interconnections and emergent processes of becoming. A genealogical sensibility recognizes that this very different view of the world is contingent and partial, and always situated within wider struggles over how to order, make sense out of, and govern social and ecological phenomena. Chapter 5 situates Holling's "very different view of the world" within the horizon opened by Herbert Simon's cybernetic thought. Situating ecological resilience theory within the wider trajectory of post-war cybernetics draws out the aesthetic dimensions of resilience. As Jean Baudrillard (1981) details, cybernetics emerged out of a modernist design aesthetic that senses, perceives, apprehends and organizes existents as rational abstractions whose essence lies in their functionality. Baudrillard locates the consolidation of this sensorial regime in the early twentieth-century creation of the Bauhaus school, but he also shows how this regime evolved into the cybernetic reduction of being to information. Teasing out these genealogical lines shows how the Bauhaus mobilized a new relation between truth and control, a will to design premised on synthetic rather than analytical reasoning. And this will to design persisted as the modernist design aesthetic rippled out from design practice to cybernetics, behavioral science, and ecological resilience theory.

Recognizing the designerly roots of resilience theory opens the question of how resilience engages with social and ecological difference. Chapter 6 explores how resilience theory transforms key elements of critique – context, conflict and subjugated knowledge – into mechanisms of cybernetic control. Resilience approaches mobilize a designerly style of critique that playfully re-combines rational abstractions to develop new systemic potentialities and capacities (Buchanan 1992; Wylant 2010). As I show through examples of traditional ecological knowledge and disaster resilience, resilience thinking mobilizes techniques of environmental power to turn any expression of socio-ecological difference into a stabilizing force. Cybernetic control emerges as resilience approaches generate difference and make difference useful to the given order of things.

Chapters 7 and 8 conclude the book with an extended consideration of the possibilities for critical thought after resilience. Resilience thinking straightjackets the possibilities for thought on complexity through the series Sustainability–Institutional Design–Feedback. That is, it structures our experience of complexity and its possible meanings in terms of the imperative to build sustainability through designing institutional arrangements that will facilitate feedback from complex systems in ways that foster social learning. In contrast, I draw from geography, anthropology, security studies, cultural studies, and radical design theory to propose a different series, Anthropocene–Hacking–Transition Design. Rather than foreclosing the meaning of interconnection and emergence, this series leaves open the possibility for more radical and subversive engagements with resilience and its cybernetic governance strategies.

The series Anthropocene–Hacking–Transition Design offers one way to visualize the present that recognizes the radical potential of contemporary social and ecological insecurities. It points towards the possibilities for inventing new ways of being-together that are sensitive not only to future systemic insecurities, but also to the various forms of suffering and insecurity that saturate the experience of the present for many humans and non-humans alike. This provides geographers and other critical scholars with a chance to reinvent critique. Rather than emphasizing the partiality of truth claims, which resilience thinking's designerly sensibilities advocate, critique can become an ethical practice, a way of engaging with a finite world that leaves open the potential to become other than we are, rather than managing the process of becoming-otherwise to ensure that nothing becomes-too-otherwise to itself as everything changes. Thus, the emergence of resilience makes critique more perilous: it heightens the chance for critique to become enrolled in cybernetic control, but it also creates the opportunity to re-invent critique in an interconnected and emergent world that has become thoroughly designed.

FURTHER READING

Excellent introductions to resilience in geography and related fields include Mark Pelling's *Adaptation to Climate Change: From Resilience to Transformation* and Kathleen Tierney's *The Social Roots of Risk: Producing Disasters, Promoting Resilience.* These volumes are particularly useful both for the detailed descriptions they provide of various debates in the literature surrounding the concept of resilience, and for the divergent slants the authors take on the politics of resilience: Pelling takes a cautious slant when he situates resilience theory in relation to broader debates over the politics of climate change adaptation, which shows how ecological framings of resilience often have conservative effects when compared to other possible approaches within the field. In contrast, Tierney is cautiously optimistic: she locates resilience as one particular way of visualizing and managing risk that emphasizes risk's social, physical and technical dimensions, which creates possibilities for enhancing social capital as a means of promoting resilience. Together, these texts provide readers with a thorough understanding of how resilience can be situated in relation to a variety of timely debates in geography. Additionally, curious readers can find excellent introductory overviews of the history, trajectory and formative debates within design studies in Nigel Cross' *Designerly Ways of Knowing* and Richard Buchanan and Victor Margolin's co-edited volume *Discovering Design: Explorations in Design Studies*.

2

RESILIENCE AS ESSENTIALLY CONTESTED CONCEPT

INTRODUCTION

One of the challenges of writing about resilience is that there is no clear definition of what resilience *is*. As we saw in the previous chapter, Brand and Jax (2007) suggest that resilience is a descriptive concept that represents some facet of reality – an inherent or learned adaptive capacity – *and* a normative concept that prescribes how human and environmental relations *should* be organized and managed, *and* a boundary object that facilitates interdisciplinary research – all at the same time. Disciplines such as geography, anthropology and international relations further muddy the water, since they are home to large contingents of critical scholars concerned with the concept's often under-acknowledged social, cultural and political effects. For example, Ben Anderson (2015) suggests that there are important differences between the way resilience functions as, alternatively, an ideology, a governmental rationality, a discourse, a metaphor, or an idiom – even if critical research tends to gloss over these different forms and produce blanket claims about resilience that lack empirical specificity. His point is that scholars indiscriminately use the term to lump together very diverse ways of knowing and governing uncertain futures. Here, governing does not refer to the practice of the institution known as "government." Instead, government refers to the conduct of conduct – it involves actions on the action of others strategically designed to produce intended effects (Foucault 2007a). As a technique or strategy of government, resilience, in all its different forms, intervenes in social and ecological processes in order to influence desired outcomes (Chandler 2014a).

As a vector for the exercise of government, resilience does not simply express normative goals. The multiple definitions of resilience, and the multiple forms each definition can take mean that resilience can express *multiple, competing* normative goals at one and the same time. Resilience can maintain the political economic status quo and ensure entrenched interests' security and persistence – a vision of resilience often found in urban security planning (Coaffee and Rogers 2008; Coaffee 2013). Resilience can advocate for transformative change in social and economic practices in order to bring about more sustainable social and ecological outcomes – a vision of resilience now commonly found amongst sustainability scientists, some geographers, and many ecologists (Folke *et al.* 2010; Wilson 2012). Resilience can cement uneven and unjust neoliberal socio-ecological relations – a vision of resilience critiqued by many critical scholars in geography and related fields (Evans and Reid 2014; MacKinnon and Derickson 2012; Welsh 2014; Grove 2014a). Resilience can subvert the political economic status quo and open spaces for performing alternative practices of state–society–environment relations (Grove 2013b; Nelson 2014). For each definition and form of resilience, there is a corresponding normative commitment and implicit ethical demand.

Writing in 1956, political philosopher W.B. Gallie coined the phrase *essentially contested concepts* to describe concepts such as freedom, democracy, sovereignty and, we could add, resilience, which lack a single, operationalized definition, and whose competing definitions carry implicit assumptions about social and political order. Essentially contested concepts are not simply terms with multiple meanings; instead, they are terms with *contested* normative implications that arose out of and thus reflect irreconcilable divisions within society (see also Connolly 1993; Walker 1993). Any effort to provide conceptual clarity to an essentially contested concept is inescapably ethical and political: it involves a choice about proper ways of living in the world and relating to other human and non-human beings; and this choice will implicitly or explicitly *negate* other possible ways of understanding and living in the world.

Studying an essentially contested concept like resilience presents unique challenges. It requires that we forego our desire for conceptual clarity and reject the impulse to operationalize a concept – for this requires coming down on one side or another (or another) in the contestation. Analysis must instead be attuned to this conflict. In our case, this means examining resilience as a *site of contestation*: one front, amongst others, in wider struggles to make sense of social and ecological experience, govern these experiences in particular ways, and resist being governed in those ways. In a provocative review essay, Stephanie Simon and Samuel Randalls (2016) suggest that researchers should direct attention to the specific *sites* in which the concept of resilience operates in order to understand what "resilience" as such actually does. That is, research should pay attention to the way resilience is mobilized in specific situations, in response to specific problems, in ways designed to produce specific social and/or ecological effects. This does not simply involve asking important questions such as "resilience of what to what" and "resilience for whom," it also requires scholars to unpack their unexplored assumptions about the form and location of resilience, and the specific effects that interventions carried out in the name of building resilience might have on collective life. This is not only a question of what resilience *does*, but also what resilience, as an essentially contested concept, *could potentially become*.

Readers familiar with the work of French historian Michel Foucault will recognize a genealogical sensibility in this style of analysis. A genealogy is at once a method and an ethic that strives to recognize contingency in social and environmental life. At its most basic, this ethic involves a sense that things could always be other than they are. The genealogist looks for continuity where others see discontinuity (and vice versa), conflict where others assert unity, contingency where others assume universality, and politics where others disavow power. Genealogy thus does not accept that terms of art or science, such as resilience, vulnerability, adaptability, sustainability or design, have coherent or transparent meaning. Instead, it asks how terms such as resilience or design have come to have the meanings, significance, and effects that they do: How, for instance, has resilience come to be understood as a system's capacity to change and adapt to emergent conditions in a way that maintains basic structure, function, and identity? How has resilience become widely recognized as a solution to individual and collective problems of living in a complex world? What kinds of interventions in social and ecological processes does an interest in building resilience enable? What are the effects of these interventions – do they reinforce the political economic status quo? Do they challenge inequalities and injustices? Do they provide new paths for subversive or transgressive action? And what possibilities does the pursuit of resilience foreclose?

A genealogical sensibility thus recognizes resilience as a site of potentiality: a site where social and ecological relations and processes gain specific meaning, but where this meaning is always exceeded by the potential for other ways of knowing, engaging with, and governing indeterminate life. This chapter introduces the key features of a genealogical reading of resilience. I begin by mapping current fault lines in debates over resilience within geography and related fields. After reviewing the various critiques scholars have advanced against resilience, I provide a brief description of what a genealogy entails, and make the case for examining an alternative genealogical tack into the concept – one that foregrounds the formative influence of a designerly ethos on early resilience research.

RESILIENCE: AN ELASTIC CONCEPT

Most accounts of resilience trace the concept's roots to discipline-specific debates in engineering, ecology and psychology (Walker and Salt 2006, 2012; O'Malley 2010; Walker and Cooper 2011; Welsh 2014). While there are important differences across all three origins (Tierney 2014), they tend to share a common concern with creating what Filippa Lentzos and Nikolas Rose (2009: 243) call a "subjective and systematic state to enable each and all to live freely and with confidence in a world of potential risks." Three examples illustrate this concern. First, in noting that resilience refers to a "world of potential risks," Lentzos and Rose indicate a specific *ontology*, or understanding of what exists, that each disciplinary understanding shares. Ecologists, psychologists and engineers all assume that the world is fundamentally unstable: it is filled with uncertainties and disturbances that cannot be prevented. In ecology, a resilient system is enmeshed in a series of complex cross-scalar interrelationships and feedback loops that generate exogenous disturbances, such as invasive species or environmental calamities. In psychology, the individual is embedded within a wider social, cultural, political and economic environment that can generate trauma, such as an abusive family member or abject poverty. In engineering, a technological system is subject to external stresses and shocks that disrupt normal systemic functioning. Despite important differences, in each case resilience involves a common relation between a system's interior and exterior: the "inside" of a system – whether this is an ecosystem, an individual psyche, or a technological system – is normally more or less stable, but is subject to unexpected disruptions from its "outside" – ecosystems at larger or smaller scales, the individual's socio-cultural setting, or the wider biophysical-technological environment. Management, design, therapy and/or engineering cannot prevent these incursions; instead, they can only prepare the system for inevitable disruptions.

Second, in noting that resilience enables "each and all to live freely and with confidence" in this unstable world, Lentzos and Rose indicate that resilience involves the capacity to withstand and potentially prosper from exogenous disruptions. According to psychologists, a resilient individual will maintain a "normal" trajectory of psychological development despite suffering trauma. Ecologists now recognize that resilient ecosystems are capable not only of recovering from disturbance, but also using the recovery to transform in ways that improve system functioning. In both cases, the concept of resilience holds out the promise of surviving and thriving even while experiencing unexpected shocks and disruptions (Aradau 2014). Similarly, engineers attempt to design resilient systems that will rapidly "bounce back" from a shock and resume normal functioning. However, there is an important difference. For ecologists and psychologists, disruption is unavoidable, and being resilient demands "permanent adaptability" to a turbulent environment (Walker and Cooper 2011: 156). For engineers, resilience is a means to *minimize* impact and thus *prevent* disturbance. The ecologist C.S. Holling (1978: 10) succinctly characterizes this engineering logic:

> Managing the distance from an undesirable region [of phase space; see Chapter 3] is within the highly responsible tradition of safety engineering, of nuclear safeguards, of environmental and health standards. It works effectively if the system is simple and known – say, the design of a bolt for an aircraft. Then the stress limits can be clearly defined, and the bolt can be crafted so that normal or even abnormal stresses can be absorbed. The goal is to minimize the probability of failure.

Thus, these visions diverge around how a resilient system should relate to an uncertain future. That the future will be unpredictable is not up for debate; instead, at issue is how individuals and collectives should understand, value, and orient themselves towards this uncertainty. For ecologists and psychologists, this uncertainty is the source of adaptive possibilities; the challenge is to devise ways of managing ecosystems and developing individual psychological capacities to both withstand disruptions and capitalize on their possibilities. For engineers, uncertainty is the source of system failure, and the challenge is to design out disruptions to the greatest extent possible.

Third, despite this divergence, in all three cases resilience refers to a capacity that management, training, and/or design can nurture (Simon and Randalls 2016). Holling's quote above illustrates how engineers attempt to design-in resilience to minimize the probability of system failure. For ecologists, many systems inherently exhibit characteristics of resilience. For example, invasive species such as the spruce budworm may prove to be highly resilient: their population fluctuates in seemingly random patterns that reflect adjustments to wider environmental conditions (Holling 1973). However, human intervention often erodes these adaptive capacities. For example, fire management in the Western US relies on strategies of suppressing fire. This reduces natural patterns of variation in the fire ecology, which typically involve periodic fires that destroy dead biotic material. While fire management reduces the number of fires that occur, the corresponding build-up of fuel means that a fire that does occur will be stronger and more damaging than it otherwise would (Holling and Meffe 1996). Given that widespread human intervention in a number of forest, wetland, riverine and aquatic ecosystems has reduced their inherent resilience, the challenge for environmental managers is often to devise ways of re-designing resilience – in artificial form – into increasingly rigid and inflexible systems. Similarly, psychologists recognize that while some individuals may be inherently more resilient than others, resilience is also a product of individuals' environments. The ability to withstand traumatic conditions and develop into psychologically "normal" adults often hinges on nothing more than the "ordinary magic" of meaningful relationships with family and the wider community (Masten 2015). Here, building resilience often involves re-introducing elements of community and sociality, such as trust and engagement, that the alienation characteristic of modern societies has eroded. Like ecology, expert psychological interventions can nurture and develop adaptive capacities to withstand traumatic shocks and stresses.

It should thus be no surprise that over the past decade, resilience has proliferated across a number of disparate fields. It responds to problems of persistent insecurity and increasingly regular surprises and shocks – problems unique to a specific experience of the world as interconnected and emergent. A world where borders that once clearly demarcated a secure inside from a threatening outside no longer hold threats at bay. A world of global environmental change and the inability of nation-states to regulate flows of atmospheric pollutants (Dalby 2002). A world of the Anthropocene, where humans have become a geological force driving planetary changes that modern science can neither predict nor control (Dalby 2013; see Chapter 7). A world of international terrorist networks and the inability of nation-states to regulate global flows of finance and people (deGoede 2012). A world of economic turbulence and the inability of nation-states to regulate financial markets and technologies that drive boom and bust cycles (Cooper 2010). A world of individual and community precarity that has eroded faith in the promise of "the good life" to come (Berlant 2011). Resilience gives a certain sense to these experiences of insecurity: it positions them as effects of complex, interconnected and contingent social and environmental milieus; and charts a course for action in the face of this uncertainty. It thus recalibrates the modernist promise of progress, growth and development into a postmodern promise that individuals and systems can survive and potentially thrive in an indeterminate world by becoming resilient (Chandler 2014a).

However, although resilience expresses a "universal mode of thinking about the relations between unpredictable subjects and their complex environments" (Dunn-Cavelty *et al.* 2015), its adoption in different fields of practice often produces contradictory forms of intervention into complex social and ecological phenomena. For example, engineering understandings of resilience as the capacity to "bounce back" and recover from shock with minimal disruption have heavily influenced the style of resilience practiced in fields such as urban security and critical infrastructure security. Here, resilience primarily involves two components: first, a "hardening" of urban space and infrastructure systems in order to make them more impervious to an attack or environmental hazard; and second, the development of emergency preparedness and response capacities in order to prevent a disruption's *effects* from causing widespread disorder. The logic here is, first, to prevent disruption from occurring, and second, to contain its impacts. Here, "resilient" cities and infrastructure allow society to quickly resume normal functioning following a disruption (see Box 2.1).

Box 2.1 URBAN RESILIENCE IN MANCHESTER

In the aftermath of the Irish Republican Army's June 1996 bombing of Manchester's city centre, city and private sector leaders intensified their ongoing redevelopment efforts. But the impact of the bombing made them consider urban design and architecture in a new light. The challenge they faced was to encourage an influx of tourists, residents and shoppers while also increasing protection for critical commercial infrastructure that facilitated these movements. They increasingly alighted on a strategy of siege-like security during large events such as the 2002 Commonwealth Games and the 2004 and 2006 Labour Party conferences. The hyper-visibility of an increased police presence, barriers and fortifications, and ubiquitous CCTV cameras were all meant to deter terrorist and criminal activities. These efforts helped usher in a renaissance of sorts for Manchester's previously run-down city centre, and helped construct an image of Manchester as a place of safety and design-led urban renewal (Coaffee and Rogers 2008).

In contrast, disaster management has long blended psychological and ecological understandings of resilience to advance an understanding of individual and community capacities required to prepare for, respond to, and recover from disaster (Brown and Westaway 2011; Tierney 2014). Since 2005, major international disaster management frameworks have made building resilience an overarching policy goal. The subtitle for the 2005 Hyogo Framework for Action (in effect from 2005–2015) was "Building the Resilience of Nations and Communities to Disasters" (UN 2005). The Sendai Framework (in effect from 2015–2030) furthers Hyogo's commitment to strengthening resilience. Their efforts to build resilience have focused on vulnerable populations' capacities to prepare for and respond to disasters, and led to a greater interest in participatory, community-based disaster management programming among international donor organizations (e.g., DFID 2011). They also focus on governance reform designed to develop institutional capacities at the national scale to support disaster risk management activities. While these developments are not without critical complications, as Chapters 5 and 6 will detail, they nonetheless illustrate how, for disaster management, resilience focuses on developing individual, community and institutional capacities to prepare for, respond to, and adapt to the constant threat of extreme events (see Box 2.2).

Box 2.2 DISASTER RESILIENCE IN RURAL JAMAICA

The community of Trinityville is tucked away in the foothills of Jamaica's Blue Mountains. A number of small villages make up the community, and for the most part, each is precariously located politically, economically, and ecologically. There is a high degree of poverty throughout the rural region, and many villages support the Jamaican Labour Party, which is the minority party in the national government but the majority party in the parish government. Moreover, many houses are located on unstable hillsides or riverbanks, which exposes residents to risks of landslides, flooding, and riverside erosion whenever heavy rains occur. Trinityville's exceptional vulnerability made it one of a handful of communities across the island that Jamaica's national disaster management agency selected to take part in a pilot program, sponsored by the Canadian International Development Agency, called the "Building Disaster Resilient Communities" project. The BDRC, as it was called, focused on developing community capacities through participatory programs that gathered and shared information horizontally between community members and vertically between communities and state agencies. However, a consistent worry was that the participatory activities struggled to attract interest from the community's youth (Grove 2013b).

However, there are conflicting uses of the term in disaster management. US emergency management places a much greater emphasis on engineering understandings of resilience, and downplays – but does not ignore – ecological and psychological understandings of the concept. In part, this reflects the unique situation in which resilience entered national popular discourse following the attacks of 11 September 2001. National leaders frequently invoked the US population's "resilience" in the wake of these catastrophic events – invocations that ironically kept the day's trauma front and center in public discourse, even as they praised the population's ability to return to everyday consumption patterns (Vale and Campanella 2005; see Box 2.3). This sense of resilience as both a systemic "bouncing back" and individual and collective coping with trauma remains strong in current US emergency management frameworks. For example, the US Federal Emergency Management Agency's recent "Whole Community" approach echoes in spirit the international disaster management community's priority on building community resilience (FEMA 2011). However, in practice, it involves a variety of participatory activities designed to align all community, local and state emergency management activities with federal emergency management goals. Of course, this reflects the contextually specific history of emergency management planning in the US federal system of government. Because American federalism grants state governments the right to govern activities not explicitly enumerated in the US Constitution, emergency management planning in the US has, since its inception in the mid-twentieth century, been riven by conflicts between federal and state emergency management goals (Roberts 2013).

Box 2.3 RESILIENT NEW YORK

re•sil•ient [re-zihl-yunt] adj.

1. *Able to bounce back after change or adversity*
2. *Capable of preparing for, responding to, and recovering from difficult conditions*

Syn.: TOUGH
See also: New York City

These definitions, found on the inside leaf of PlaNYC's 2013 *A Stronger, More Resilient New York*, provide a pithy definition of resilience for the uninitiated. The document itself was a product of the New York City Special Initiative for Rebuilding and Resilience, formed by Mayor Michael Bloomberg in December 2012 following Superstorm Sandy. Sandy was one of the costliest storm events on record in the United States, causing US$19 billion in damages in New York City, and US$65 billion across the eastern Atlantic seaboard. But despite these losses, in the aftermath, emergency management professionals focused on how Sandy could have been worse. Had the storm struck at high tide, or made landfall further north, the storm surge would have impacted Hunts Point Food Distribution Center in the Bronx, impacting much of the city's food supply.

There was a general recognition in after-action reports conducted by both the NYC Office of Emergency Management and the Federal Emergency Management Agency that despite a few setbacks and challenges here and there, by and large the emergency response was effective. There was no cataclysmic social breakdown, vital services and utilities were (for the most part) restored in a timely manner, and the forty-three deaths, while tragic, were a remarkably low number for a storm of Sandy's magnitude hitting one of the world's largest cities. As a result, both after-action reports focused on the need to better prepare for a storm that could be worse than Sandy.

But the synonym and alternative definition also tie in the city's resilience-building efforts with longstanding imagery of New York's recovery, strength and toughness that circulated in the months and years after 11 September 2001. More than a series of recommendations on infrastructure protection, *A Stronger, More Resilient New York* framed the city as persevering in the face of growing threats from climate change – a perseverance that could be strengthened by following the plan's recommendations for increasing the resilience of the city's critical infrastructure systems and communities (City of New York 2013).

The list could include examples from other fields such as climate change adaptation (Eakin and Leurs 2006; Nelson *et al.* 2007; Pelling 2010), local and regional economic development (Welsh 2014), or international peacekeeping and humanitarianism (Duffield 2012; Chandler 2014a). Many critical and applied scholars have noted this conceptual fuzziness, but the key point here is that these definitions are often diametrically opposed to one another. Consider urban resilience. As Box 2.1 and 2.3 illustrated, resilience can refer to the elimination of a threat and engaging in an effective disaster response. But as Mike Hodson and Simon Marvin's (2009) discussion of urban ecological security shows, resilience can also refer to urban growth strategies designed to maintain and even strengthen local economic development and urban competitiveness in a dynamic global economy that climate change makes even more precarious. Urban resilience can *also* refer to new ways of approaching urban environmental management and sustainability planning that visualize cities as complex ecosystems (Grimm *et al.* 2000). And urban resilience can *also* refer to urban design and planning initiatives that seek to enhance quality of life and make cities more livable in the face of uncertain environmental and economic futures (Washburn 2015; Cohen 2016). To be sure, quality of life is often an important component of urban growth strategies and urban competitiveness. But it nonetheless indicates that "urban resilience" can refer to opposing ways of knowing and governing uncertain social and ecological processes at one and the same time: if urban design mobilizes a vision of resilience that enables urban residents to *live with* and *accommodate* threats such as rising sea levels and increasingly frequent floods, urban security makes resilience a matter of *preventing* threats from materializing in the first place.

The relation between resilience and risk provides another example of the concept's incoherence. Many scholars frame resilience as the converse of risk. They see risk as a passive condition that involves exposure to a hazard coupled with susceptibility to loss, or vulnerability (see Wamsler 2009; Beilin 2015). This view of risk has been influential within the field of disaster management (Wisner *et al.* 2004). It has resulted in disaster risk management and mitigation policies and initiatives that emphasize expert, technocratic interventions designed to reduce vulnerable peoples' exposure. In contrast, resilience emphasizes vulnerable peoples' self-help capacities and empowers them to engage in their own disaster mitigation programming (Brown and Westaway 2011; but see Grove 2014a). If risk induces anxiety about an uncertain future, resilience offers optimism. The critical international relations theorist Mark Neocleous (2013: 3) characterizes this argument thusly: "rather than speaking of fragility and its (negative) associations, we should be speaking of resilience and its (positive) connotations." However, this vision contrasts markedly with critical scholars' assertions that resilience is simply the latest fold in an ongoing politics of security. Here, resilience creates an atmosphere of *constant* vigilance and risk. It portends a future of inevitable disruption. It denies the possibility of safety from harm, and as Neocleous notes, thus "subsum[es] and surpass[es] the logic of security" (ibid.). Rather than the converse of risk, for these scholars, resilience *amplifies* risk and makes risk an inescapable part of everyday life. It denies the possibility of safety from harm, and forces individuals and collectives to *embrace* rather than avoid risk as the only means of surviving the turbulent future (Coaffee *et al.* 2009; Evans and Reid 2014).

Resilience thus appears to be an infinitely elastic concept. It can be bent, twisted, folded in on itself, and shaped into all sorts of different forms that bear little resemblance to one another. For some proponents, this makes resilience an attractive boundary object that can facilitate transdisciplinary work on the dynamics of complex social, technological and/or ecological systems (Brand and Jax 2007). It can bridge otherwise disconnected research in fields such as ecology, engineering, geography, sociology, architecture, public administration and urban planning. It thus introduces a postmodern research paradigm that does away with artificial disciplinary divisions. But at the same time, this malleability allows different interest groups to co-opt resilience within a variety of contradictory and even conflicting political projects (Brown 2013). It can at once be a vehicle for pragmatic change, potentially subversive community activism *and* a mechanism that solidifies uneven political economic and ecological relations (Grove 2013b). Indeed, even as it promises positive, progressive and transformative change, in practice resilience programming has often had conservative and reactionary effects that preserve the political economic status quo against the social, economic and environmental uncertainties it created in the first place (Pelling 2010; MacKinnon and Derickson 2012). Critics of resilience have honed in on these reactionary effects. The next section explores some of these critiques through examining debates in disaster management.

RESILIENCE AND ITS DISCONTENTS

When the 2005 Hyogo Framework made building community and national resilience to disasters its overarching goal, many disasters scholars questioned what, if anything, a focus on resilience brought to the field (Manyena 2006). Disaster researchers had been examining many of the same processes of vulnerability, adaptability, and socio-ecological change that interest resilience scholars for a number of years, but through a different conceptual and theoretical framework. However, other scholars advanced a more pointed critique: that resilience diverts attention and resources away from the root causes of vulnerability (Cannon and Müller-Mahn 2010; Gaillard 2010). At issue here is how resilience approaches downplay structural determinants of vulnerability and emphasize self-help capacities. Since the 1970s, radical disaster studies has detailed how vulnerability is a product of uneven capitalist political economic relations that create marginalized populations and force these populations onto marginalized lands that are highly exposed to environmental hazards (O'Keefe *et al.* 1976; Hewitt 1983; Watts 1983). Effectively addressing the human and economic costs of disasters requires identifying and addressing these uneven political economic relations as they play out across scales from the local to the global (Maskrey 1989, 1994). It also requires a focus on pre-event risk mitigation efforts rather than developing post-event response and preparedness capacities. However, the recent adoption of resilience as an overarching policy goal in the Hyogo and Sendai Frameworks has increasingly emphasized developing individual and community capacities to live with vulnerabilities. This is part of a wider turn towards "therapeutic" forms of development throughout the international community. Rather than holding out the modernist promise of eradicating inequality, since the late 1990s, development programming has relied on participatory initiatives to manage the effects of poverty and inequality *in situ* (Pupavac 2001; Duffield 2008). These programs attempt to moderate marginalized peoples' expectations for a better future and develop social bonds to create self-reliant communities. Disaster management has recently adopted many of these strategies and techniques as part of an effort to build resilient "cultures of safety" that take responsibility for mitigating their exposure and managing the initial stages of disaster response (Grove 2013a; see Box 2.2).

Critical disaster scholars have advanced two important critiques of this resilience turn. First, "therapeutic" styles of disaster management are arguably less effective. On one front, they divert scarce and financial resources away from hard infrastructure improvement projects such as shoring up riverbanks with Gavian baskets, clearing gullies that flood, or building check dams that can halt a landslide and towards community-building activities that are not as effective at mitigating physical risks (Cannon and Müller-Mahn 2010). On another front, they divert attention away from the real drivers of vulnerability. The problem here is that disaster resilience re-conceptualizes vulnerability as a problem of a system's *internal* characteristics. It does not view vulnerability as an effect of wider structural inequalities that marginalize certain peoples and places and make them more susceptible to loss, nor is it a matter of geographically uneven development that produces underdeveloped states that lack resources to support effective risk mitigation and disaster planning efforts. Rather, vulnerability is an effect of cultural and institutional constraints that limit individual and community adaptability.

This vision of vulnerability is an outgrowth of resilience proponents' tendency to adopt a *functionalist* view of society. Chapter 4 will explore some reasons for this; for now, we can note that functionalism assumes that society has an objective reality and is inherently cohesive, and that it comprises interconnected parts that *function* as a system to maintain equilibrium and stability for the whole. As a result, it evaluates social processes from a teleological perspective of how they contribute to social stability. These functionalist assumptions inform resilience thinking's emphasis on the cultural and institutional limits to individual and community adaptability: vulnerability is a sign of a social system that has been thrown off of its equilibrium; these distortions can be adjusted through expert interventions that re-calibrate both culture and institutions in order to improve adaptability and ensure society as a whole can remain stable in a world of flux and change. Thus, a functionalist solution to vulnerability emphasizes governance reform and participatory interventions designed to bring about cultural change – precisely the therapeutic interventions disaster resilience champions. Governance reform, according to proponents, will increase state capacity to plan for and manage disaster events. It will also increase individual and community trust in government institutions. Engineering cultural change, in turn, will create those self-organizing "cultures of safety" that utilize local knowledge and experience to manage disasters without relying on aid from the state or international donors (Grove 2014a).

In this functionalist reading, vulnerability is a problem because of *internal* conditions: the lack of adaptive capacity or flawed institutional or cultural practices leave individuals, communities and nations unable to properly adapt to emergent futures. Wider structural conditions do not enter the equation; they are part of an external world individuals and collectives cannot control but can only adapt to in more or less appropriate ways (Chandler 2014a; Evans and Reid 2014). This vision view of vulnerability leads into a second critique of the resilience turn in disaster management: its tendency to downplay questions of politics and power relations. Terry Cannon and Detlef Müller-Mahn (2010: fn5) succinctly detail this issue:

> When Folke et al. [in their 2005 paper on adaptive management and environmental governance] discuss the influence of people-centered

> involvement in the governance of social-economic systems in this substantial paper, it reflects an assumption that powerful actors will respond to ecosystem problems by engaging in co-management (as in adaptive governance) without much consideration of self-interest and power relations that have damaged the ecosystem in the first place.

At issue here is how much resilience scholarship assumes that "resilience" is, first, a desired condition, and second, that there is consensus on this desired state. Of course, this reflects the functionalist assumption that society is inherently coherent and founded on consensus. But as Cannon and Müller-Mahn (ibid.) go on to note, human systems "embody power relations and do not involve analogies of being self-regulating or 'rational.'" Assuming there is consensus erases potential dissensus and overlooks both conflicts over resources and the constitutive role uneven power relations play in socio-ecological change (Brown 2013).

More importantly, assuming consensus also blinds resilience scholarship to the possibility that it might unintentionally shore up the social, economic and environmental insecurities it purports to solve. At issue in this latter point is how resilience approaches normalize insecurity and vulnerability. Critical scholars have shown how the growing interest in resilience goes hand in hand with the rise of neoliberal political economic reforms. Since the late 1970s, neoliberal governance strategies have dismantled the Keynesian Welfare State in the name of combating economic stagnation (Harvey 2005; Klein 2007). Neoliberal reforms have drastically reduced social services such as unemployment compensation and job retraining programs, welfare support for un- and under-employed workers, and social security payments that provide the elderly with financial stability – to say nothing of labor regulation reforms that have curtailed union influence and environmental regulation reforms that have increased pollution. These reforms have gone hand in hand with revanchist policing strategies that have heightened the risk of violence and incarceration for poor and minority populations (MacLeod 2002), as well as draconian cuts to public education and health provisioning that improve a population's quality of life and life chances. The result is heightened precarity and insecurity – economic, psychological and physical – for many individuals and communities (Berlant 2011). Resilience enters the scene as a strategy to live with these insecurities. As we saw above, it does not promise safety from danger and threat, but rather asserts that becoming resilient will enable individuals to survive and thrive in unstable conditions. Resilience thus normalizes and depoliticizes vulnerability: it asserts that insecurity and instability are a natural part of living in a complex, interconnected and emergent world, rather than the result of historically specific political economic practices (Duffield 2011; Evans and Reid 2014).

For other critics, resilience is an intensified form of *securitization*. This refers to the process of framing and managing various social and ecological phenomena through the language of security. As scholars in critical geopolitics and critical security studies have shown, invocations of security, threat, danger, insecurity, and so forth are political resources that instantiate power relations, construct identities, and legitimize particular forms of socio-ecological order by delineating who or what is threatened, and thus in need of security, and who or what is threatening, and thus in need of surveillance and regulation (Campbell 1992; Dalby 2002; Grove 2010). The adoption of resilience thinking in fields such as urban planning, public health, finance, disaster management, and capital risk management has opened novel fronts in the politics of security (Neocleous 2013; Coaffee 2013). Resilience approaches render everyday activities and banal social and environmental relations security concerns, extending a politics of security to the spaces and relations of everyday life and creating anxiety-ridden atmospheres of vigilance, risk and inevitable crisis (Cretney 2014; Grove 2014a).

Taken together, these critiques have raised the importance of asking critical questions such as, "resilience of what, to what and for whom?" (Cretney 2014). To their credit, many resilience scholars who have played an influential role in propagating the "gospel of resilience" (cf. Nadasdy 2007) have attempted to engage with and address some of these concerns (Folke *et al.* 2010; Walker and Salt 2012). Indeed, much contemporary work now revolves around the problem of how to integrate resilience thinking's pragmatic slant on environmental management with research approaches more sensitive to resilience's normative dimensions and power dynamics (Brand and Jax 2007; Miller *et al.* 2008; Miller *et al.* 2014). For example, some resilience thinkers have argued that accounting for the impact governance arrangements have on adaptive capacity can effectively incorporate politics into studies of complex social and ecological dynamics (Folke *et al.* 2005; Lebel *et al.* 2006; Folke 2006). Recently, this line of thinking has also begun to draw on transitions management literature in order to understand how to manage processes of institutional change (Folke *et al.* 2010). Transitions management recognizes that resilience may be undesirable if a given system is unsustainable. In such cases, systemic transformation may be required, such as a shift in production systems or the adoption of new cultural norms and values (Rip and Kemp 1998; Kemp and Rotmans 2005). Here, politics can inhibit transformation in two ways: existing political institutions can be rigid and inflexible, preventing necessary adaptations; and competing visions of sustainability and resilience reflect entrenched social and political economic interests (see Grove 2013b; also Chapter 6). The challenge for transitions management is to devise ways of overcoming these limitations, such as adaptive management, reflexive governance, and participatory initiatives that encourage collaborative learning (see Chapter 4). Still other resilience proponents have attempted to make room for competing understandings of resilience. This is the goal of Brand and Jax's (2007) distinction between resilience as a descriptive concept, normative concept, and boundary object, discussed above in Chapter 1. As a descriptive concept, resilience has a clear, objective and operationalizable definition for ecologists studying ecosystem dynamics. As a boundary object, resilience facilitates interdisciplinary research into complex social and ecological systems and their management. The latter recognizes how resilience functions as a normative concept that implicitly conveys subjective value judgments on social and ecological processes. As a boundary object, resilience can facilitate cross-boundary deliberations over "who resilience is for" and "what we need to be resilient against."

These efforts to accommodate criticisms indicate many resilience scholars' willingness to take seriously critics' concerns. However, to a certain extent, these accommodations overlook how many critics call into question the demand to become resilient in the first place. For example, critical security studies research asks how it is that some issues – such as social and ecological complexity, or individual or community adaptability – have become named as security problems in the first place, and unpacks the power effects of this naming. Here, the issue is not only about subjective questions of "resilience of what, to what and for whom;" it is also a matter of *who decides* on the need for resilience in the first place. This is a much more destabilizing question: rather than taking for granted the need to become resilient and deliberating over the best way to develop resilience (see Box 2.4), this critique calls into question the demand for resilience on ethical and political grounds. Critics may propose alternative slants on resilience, perhaps grounded on ideas of transition (Wilson 2012), subversion (Grove 2013b), or against everyday violence (Scheper-Hughes 2008), or they might propose alternative ways to address social and ecological insecurities, such as vulnerability (Cannon and Müller-Mahn 2010) or resourcefulness (MacKinnon and Derickson 2012). And they may attach wildly different normative values to each form of (anti-)resilience, from the dominant vision's maintenance of the status quo and transformation's emphasis on pragmatic reform, to subversion or everyday resilience's interest in our ethical responsibility to attend to marginalized peoples' struggles to get by in enervating and unjust social and ecological worlds.

Box 2.4 POLICING RESILIENCE

Brand and Jax's (2007) effort to distinguish between resilience as a descriptive concept and a boundary object is a notable effort to account for polarizing debates on the use and potential shortcomings of resilience. It has spurred new lines of questioning and opened new possibilities for provocative collaborations across the social and natural sciences, engineering, design and architecture (Miller *et al.* 2008). However, to a certain extent, it also reinforces polarization between the sciences. We might read their attempt to corral criticisms of resilience within the conceptual "space" of a subjective boundary object as a rear-guard strategy to preserve the continued prominence of ecological understandings of resilience – precisely the prominence that critiques have called into question over the past decade (Cannon and Müller-Mahn 2010; Davidson 2010). From this angle, Brand and Jax's argument is a fascinating example of *policing* disciplinary boundaries. The term *policing* draws on philosopher Jacques Rancière's differentiation of *police* and *politics*. For Rancière (1995, 2001), police roughly signals practices of boundary creation and enforcement that stabilize a given order of things. Politics, in contrast, refers to an act that transgresses these boundaries and introduces novelty – something that cannot be accounted for within a given "police" distribution. If we take Rancière's understanding of "police" to mean an act that assigns something to a specific space and time in order to give it a specific identity, function, or role, then Brand and Jax provide an excellent case of disciplinary policing. Their differentiation of resilience into a descriptive concept and boundary object erects a scaffolding of sorts, a structured series of inclusions and exclusions that clearly demarcates the proper roles for ecologists, geographers, economists, and other social scientists interested in complex systems and resilience: ecologists are keepers and refiners of the concept's descriptive content; all others may utilize this concept to clarify their analyses of socio-ecological systems and develop policy interventions on the basis of these analyses. An implicit hierarchy remains in place. Brand and Jax's argument appears here as an attempt to *fix* or *stabilize* the significance of resilience in a way that removes contestation over the term's utility for understanding social and ecological change. While there may be subjective debates over whether or not resilience for specific systems is desirable or undesirable (for example, capitalist socio-economic systems with high degrees of income and racial inequality), what is not up for debate is resilience's utility for understanding how these desirable or undesirable changes may occur. Brand and Jax's understanding of resilience as a boundary object creates a space for ecologists, geographers, economists, and other social and environmental scientists to debate *how* resilience can inform interdisciplinary research on social and ecological systems, but it does not allow for researchers to question whether or not the concept of resilience and its corresponding systems ontology *should* be used to inform research on social and ecological change.

The meaning of resilience is thus neither transparent nor objectively determined. Rather, it is essentially contested (Gallie 1956; Connolly 1993): it is bound up in ongoing debates and struggles over how to live in a world without the guarantees of modern security. In a world where borders and boundaries do not corral spatially interconnected threats, how should we relate to both near and distant humans and non-humans? In a world where prediction cannot tame emergent futures, how should we relate to the immanent potential to change, to become other than we are – both as individuals and (social and ecological) collectives? These questions have no straightforward answers; instead they are undecidable ethical and political aporias that open onto a multiplicity of distinct possible forms of life.[1] As political theorists have documented through examples of other essentially contested concepts such as democracy, freedom and sovereignty, attempts to assert one definition of an essentially contested concept over others are political acts that foreclose this aporia and negate other possible forms of life (Gallie 1956; Connolly 1993; Walker 1993). Any attempt to give resilience determined meaning – whether as descriptive concept, boundary object, normative concept, ideology, discourse, metaphor, and so on – is likewise an unavoidably political act.

Given the politically charged nature of the concept, my concern is not to try and clarify what resilience is, but rather to take this contestation itself as an object of study, and unpack how resilience has become the essentially contested concept it is today. This question calls for a *genealogical analysis* of resilience. As the introduction to this chapter hinted, a genealogy is both a mode of analysis and an ethic attuned to the contingent and conflictual in the present. Because it strives to analyze the broader conflicts, strategies and struggles that attempt to stabilize contested concepts, it is uniquely suited to examine an essentially contested concept such as resilience. The next section details what a genealogical analysis entails.

TO DIE OR NOT TO DIE: GENEALOGIES OF RESILIENCE

Because our understandings of genealogy have been significantly shaped by the writings of Michel Foucault, understanding what these features of genealogy involve requires us to shift gears and directly engage with his work. Foucault (1977a, 1977b) developed his genealogical style in order to unpack how the foundation of European modernity – the sovereign, self-writing subject (in Descartes' famous formulation: "I think, therefore I am") – was historically constituted across a variety of seemingly disconnected practices, including carceral punishment, clinical practice, psychological confinement, and transformations in the life sciences and humanities. This enabled Foucault to identify the subject as a contingent accomplishment and an effect of power relations rather than a timeless essence that could guarantee truth and security. Genealogical inquiry thus complexifies rather than clarifies an analytical field: it confronts claims to universality with the empirical residue of ongoing struggle over how to govern and be governed, and thus shows that things could always be other than they are. As political theorist Wendy Brown (1998: 42) describes it, to engage in genealogical thought is to "identify the discourses and political rationalities constitutive of our time such that they are brought into relief as historical, contingent, partial, and thus malleable, such that 'that-which-is' can be thought as 'that which-might-not-be.'"

Critical scholars have mobilized this genealogical sensibility to unpack how resilience has become an influential solution to problems of complexity and emergence. For some, this analysis suggests that resilience is ideologically compromised, nothing more than an extension of the ongoing neoliberalization of social and ecological relations; while for others, the concept retains transformative and even subversive potential. The discussion that follows explores some key features of genealogical thought and how they have influenced critical work on resilience. While by no means exhaustive (see Lemert and Gillan 1982; Philo 2007; Elden 2001; Johnson 2008; Collier 2009; and Dittmer 2010 for further discussions), these features include:

- Genealogy is a form of *spatial analysis;*
- The spaces genealogy addresses are *beatified spaces*: dynamic sites of conflict and contemporaneous juxtapositions that are always on the verge of becoming other than they are;
- Genealogy offers a *bellicose history*: a style of history that seeks to identify conflict, contingency, and contemporaneous multiplicity rather than uncover origins.

Genealogy as spatial analysis

A genealogy is first and foremost a kind of *spatial analysis.* Over the years, several geographers have drawn out the spatial dimensions of Foucault's thought (Philo 1992, 2007; Elden 2001; Johnson 2008). His analyses of asylums, prisons, clinics and sexuality – to name but a few – revolve around exploring *spaces of dispersion*, or the series of relationships between concepts that give each their meaning. From a genealogical perspective, words have no deep or essential meaning. Meaning does not follow the structure of representation, which assumes that a signifier (word) corresponds to a signified (thing, concept, sensation, phenomenon). Indeed, Foucault (1994) argues that there is always more to say about a given situation than what we are able to put into language. We might recognize this in our own lives whenever we come across situations that we "just can't find the rights words for." The frustration we feel at not being able to put all our thoughts, feelings, sensations, and experiences into language expresses one of the fundamental tenets of post-structural thought many associate genealogy with: there is always an excess of "life" beyond language.

Foucault derived his particular understanding of meaning in terms of spaces of dispersion from careful study of language and meaning-making in experimental literature. In *Death and the Labyrinth* (Foucault 2004), Foucault explores how the writer Raymond Roussel stretched language to its breaking point. Reading Roussel's *La Vue*, Foucault (2004: 109) suggests that Roussel "opens a universe without perspective." Here, Foucault details Roussel's attempts to describe a scene without a privileged point of perspective. This requires Roussel to utilize a "system of directions of proximity passing from one to the other as if following the links in a chain" (ibid.). For Roussel, these directions are not relative to the viewing subject, but instead are relative only to the specific element being described. There is thus no "depth" to the scene. There is no foreground that prioritizes certain things over those in the background. Instead, he very simply attempts to describe what exists in the scene by describing, in the finest level of detail possible, how everything relates to everything else. But as Foucault notes, this venture fails, for there is no end of things to describe when there is no privileged viewing perspective that affords certain things more importance over others: "this inexhaustible wealth of visible things has the property… of parading in an endless line… It always shows something else asking to be seen; there's no end to it" (ibid.: 112). This endless chain of relations that cannot be fully encapsulated in language is what gives meaning to the elements within the scene. In Foucault's reading, Roussel *spatializes* meaning: the meaning of things depends on nothing more or less than the (spatialized) *relations* between different things.

Chris Philo (1992) and Peter Johnson (2008) have both emphasized the importance of these "spatializations of literature" in Foucault's thought: Foucault turned this interstitial understanding of meaning towards the historical archive, in order to analyze how taken for granted concepts such as "madness," "health," "prison," "punishment" and "sexuality" derive their meaning from relations with other terms such as rationality, freedom, the family; each of which derives its meaning from relations with other terms, and so on. More recently, scholars have extended this relational understanding of meaning to resilience, and thus have explored how the meaning of resilience hinges on relations with other concepts such as uncertainty, complexity, governance and life, each of which takes their meaning from relations with other concepts, and so forth. This work details how our current understandings of resilience have been shaped by debates and developments in seemingly unconnected fields such as economics, biology and military planning. Jeremy Walker and Melinda Cooper's (2011) influential genealogy shows how C.S. Holling's ecological understanding of resilience and Austrian economist Friedrich Hayek's neoliberal economic thought share a common ontology of complex systems. Their understanding of complexity hinges on a set of spatialized relations between life, society, the state, planning, nature, economy, adaptation, vulnerability, governance and development. Complex systems are, by their very nature, dynamic and self-organizing, and this organization is driven *not* by an environmental or economic planner who stands outside the system and controls the system, but rather through innumerable feedback loops between human, non-human, institutional and technological elements *within* the system. To exist in a world of complexity is thus to be ontologically vulnerable – "complex" life is always vulnerable to potential disruptions, because it exists as such on the basis of complex interrelations that could change on their own. In such conditions, centralized planning's efforts to optimize system performance through centralized control and top-down management (more on this in Chapter 3) will inevitably create rigidity and immobility where flexibility and adaptability are required. Adaptability thus becomes the means of survival in a vulnerable world. In order to grow, develop, and enhance well-being in a complex world, governance now needs to enhance adaptability through decentralizing decision-making authority and implementing reflexive adaptive management techniques. In Hayek's case, this involves a neoliberal agenda of "rolling back" the welfare state in order to bring about a free market populated by individual decision-makers. In Holling's case, this involves resilience-building initiatives based on principles of adaptive management, which Chapter 3 will detail.

Brad Evans and Julian Reid's (2014) genealogy adds another layer to Walker and Cooper's analysis. They detail how ecological resilience refers to a particular kind of adaptive capacity: living organisms' capacities to respond to changes in their complex and emergent environments. They trace the roots of this assertion that complex life is unavoidably vulnerable to arguments in biology and ecology that becoming adaptive requires exposure to danger and harm. Their work situates formative understandings of resilience not only within neoliberal economic theory, but also in relation to understandings of life, complexity and adaptability that developed in early to mid-twentieth-century life sciences.

David Chandler's (2014a) genealogy of resilience muddies this picture. Chandler carefully distinguishes the forms of complexity that underpin neoliberal economic thought and ecological understandings of resilience. For Chandler, neoliberal thought refers to what he calls *simple complexity*. Simple complexity recognizes that systems will exhibit unpredictable, non-linear behavior, but asserts that the system is closed. This means there is always the possibility of an external or transcendental position from which a system can be known, judged and controlled. For neoliberal economists, market rationality is one such transcendental determinant of system behavior. This external position provides a foothold for experts to control the system: because the market determines system behavior, management interventions are required to ensure that market forces operate without distortion. This vision of simple complexity contrasts with what Chandler calls the *general complexity* of resilience thinking. General complexity asserts that there is no external vantage point from which system dynamics might be known and controlled. Rather than positing a topographical division between system and world, it asserts a topological relation between the system and its outside, in which the system is always already part of the world. There is thus no external position that allows for absolute knowledge and control. Instead, governance must constantly reflect on and respond to life's creative, emergent powers. Resilience thinking does not force complex life to conform to externally imposed policy agendas determined by market logics; instead, it requires limitless governance in order to facilitate, rather than control, "natural" life processes.

In their own ways, these genealogies show that the meaning of resilience cannot be reduced to disciplinary debates in ecology, psychology and engineering. Instead, meaning emerges out of contingent alignments between a variety of different concepts that developed as people tried to make sense out of new experiences such as biological evolution or economic depression. While some critics may read this as a sign that resilience is inextricably bound up with neoliberal governance reforms, Chandler's analysis in particular emphasizes the contingency of any alignment between resilience and neoliberalism. In his reading, resilience is a "resonance machine," a site where multiple historical trajectories are superimposed on one another in ways that can amplify certain tendencies, such as a shared critique of top-down, linear approaches to governance. But his emphasis on resonance recognizes that there is no essential connection between resilience thinking and neoliberal governance. Any affinities are contingently worked out in particular contexts, in response to specific problems of government, and can potentially be undone. As a resonance machine, resilience can take on different meanings in different contexts, and align with different political projects as its meaning and political significance shift. The question then becomes how these changes occur – a question the next subsection explores through examining the *beatified* qualities of Foucault's spaces of dispersion.

Beatified space

Foucault's use of space has been the focal point of critiques from some geographic quarters. For example, David Harvey (2007) takes Foucault to task for relying on a Kantian understanding of space that ignores how space is relationally constituted. For Harvey, Foucault's spatializations are a form of abstract, absolute space – a two-dimensional surface on which the genealogist can draw chains of relation between different elements. But Harvey passes over a second characteristic of genealogical thought that complicates his critique: the spaces of dispersion that constitute meaning are never fixed and static, but are instead *dynamic sites of conflict and juxtapositions that are always on the verge of becoming other than they are*. The space of dispersion is thus not an absolute space as much as it is a *beatified space* (Agamben 1990): a space that shimmers with potential to become other than it is, potentiality that results from the specific ways words and things are uneasily juxtaposed with one another.

Gilles Deleuze's discussion of Foucault's *oeuvre* helps clarify this second characteristic of genealogy. Deleuze (1988) uses the geological term "strata" as a metaphor to signal the sedimented series of relations that align words with things and thus make meaning possible. This metaphor works on multiple levels. On one level, it refers to a sort of internal consistency: just as earthen materials that comprise a given stratification share a common series of traits and characteristics, so too are the spaces of dispersion Foucault charts internally uniform: within a given historical and geographic period, meaning – the contingent relations between words and things – remains more or less constant. On another level, strata can be differentiated from other strata. Different strata represent distinct historical periods. Similarly, Foucault approaches history not as a search for origins, but rather as an exploration of shifting strata. As Lemert and Gillan (1982) note, this understanding of history reveals Foucault's debt to the *Annales* school of history in France, and particularly Fernand Braudel's work, which was influential in France during the mid-twentieth century (see also Foucault 1972). In brief, *Annales* historians saw history as marked by discontinuous breaks driven by slowly changing process (such as changes in everyday activities and diet; Braudel 1992) that eventually built up to a threshold beyond which established institutions, beliefs, practices, and meanings no longer held, and new ones took their place. Each period can be thought as a stratification in the historical archive – a plane (literally) of consistency distinct from its predecessors and antecedents.

However, on a third level, Deleuze's metaphor also points to the way that there is always something *outside* or *beyond* strata – a realm of the *non-stratified*. In geology, these are geochemical, biochemical, or climatological forces (to name a few) that shape the consistencies found within a stratum. This exterior makes stratification possible – these are the forces that bring materials into contact and compose, or more precisely, assemble strata into coherent planes differentiated from one another. But they also reveal the contingency of stratification, and the potential for strata to become different. These external forces are forces of beatification – they express the potential to become otherwise. Deleuze suggests that for Foucault, force also shapes stratified spaces of dispersion – but these are force *relations*: conflicts, tensions, juxtapositions, and battles between opposing factions within society. Meaning hinges on nothing more or less than the outcome of conflicts that stabilize relations between words and things.

In the case of resilience, as we have already seen, the evident variety of normative goals that are attached to different forms of resilience suggests that this conflict is active and ongoing. From a genealogical perspective, the inability to identify a single, authoritative definition of resilience indicates how the concept is caught up in a number of different struggles over how to understand, organize, and manage social and ecological relations in a complex, indeterminate world. *Resilience is thus a beatified concept*: it overflows with the potential for our current understandings of social and ecological relations to become other than they are. In this sense, a genealogy is less concerned with what resilience is or *should be* than with understanding these potentialities – and the force relations that have shaped these potentialities. This is a question of the *bellicose history* of resilience.

Bellicose history

The beatified spaces of genealogy are spaces that express the potentiality to become otherwise because they are never unitary, whole or stable – they are spaces of conflict, juxtaposition, tension and contemporaneous multiplicity (cf. Massey 2005). Different thinkers give different names to the potential to become otherwise – the potential for people, things, groups, processes, and so forth to change form or function, take on new identities, meanings, value, and so forth: for Deleuze, the virtual; for Bataille, the outside; for Nietzsche, desire; for Derrida, *diferánce*; and so on. Foucault names this outside *warfare* (see especially Foucault 1998a, 1998b, 2003). To be sure, some commentators read his use of warfare as a coarse and naïve understanding of power in terms of domination and oppression – one he quickly cast aside in favor of more nuanced studies of governmental rationalities and practices (Pasquino 1993; Dean 2004). However, I follow Lemert and Gillan (1982) and Philo (1992, 2007) in understanding warfare as a marker for the contingency of the present and the immeasurable potential for things to be other than they are.

Chris Philo's (2007) essay on Foucault's engagement with discourses of warfare in his 1975–1976 Collège de France lectures (Foucault 2003) is particularly informative. Philo convincingly argues for understanding warfare as a crucial foundation of sorts for Foucault's genealogical thought. In his reading, warfare provides an imagery and language that allow Foucault to explore the conflictual and contingent emergence of political rationalities. Key here is Foucault's (2003) elaboration of *subjugated knowledges*. Subjugated knowledges are essentially disqualified forms of knowledge that remain below the threshold of discursivity. Foucault identifies two forms of subjugated knowledge: First, there are "historical contents," the experience of conflict and struggle masked by totalizing systems of knowledge. Second, there are the disqualified forms of knowledge excluded for being insufficiently "scientific" or objective. The former are the raw materials of history, the everyday combats of the past that shaped the present array of forces; the latter are interpretations of these raw materials. Together, these two forms of subjugated knowledge make possible an alternative interpretation of liberal social order, which disrupts totalizing narratives that celebrate the progressive unfolding of universal human rationality with the messiness, contingency, conflict and – importantly – potential to become otherwise of localized battles.

Genealogy identifies subjugated knowledge and brings this conflict to the fore in order to destabilize the given order of things. Subjugated knowledges are the "remains" of battle; they express the possibility that things could have been other than they are, and thus demonstrate the contingency of the present. This brings me to the third characteristic of genealogical thought: it produces a specific style of historical analysis, what Lemert and Gillan (1982) call a *bellicose history*. We can usefully contrast bellicose history with conventional historical analysis, or what Foucault (1972) calls a "total history." Total history is a search for coherence, essence, deep meaning, and origins. It attempts to impose a unified, *a priori* historical narrative that smoothes out confusion, contradiction and contingency on the ground. As Foucault (ibid.: 10) writes, "a total description draws all phenomena around a single centre – a principle, a meaning, a spirit, a world-view, and overall shape." Total history thus offers coherent narratives of progressive change that read past events in terms of their relevance for our contemporary present.

In contrast to total history, Foucault's mode of historical analysis uses history to destabilize the present and show that what we take for granted is itself a contingent product of historical struggle. This is the crux of bellicose history: to show that conflict, division, and contingency exists where we commonly assume coherence, essence, and permanence. History for Foucault is not a process of progressive, teleological social change inevitably leading to the betterment of humanity, but is rather "rearrangements in the relations among the multiple forces – material, economic, social – that comprise a social formation" (Lemert and Gillan 1982: 12). Bellicose history enables Foucault to explain concrete practices in a way that does not rely on an interpretive history that searches for deep meaning devoid of context, but rather "maintains the concrete struggle of history" (ibid.: 33). History has no deep or essential meaning; instead, history is intelligible in terms of struggles, strategies and tactics, and the task of genealogy is to bring these constitutive conflicts to light.

As a method of bellicose history, genealogy thus engages in a dual maneuver: in one move, it *de-composes* the linear, progressive series along which total history has organized historical events. In a second move, genealogy *re-composes* these events along a different series, a series of conflicts and battles that reveal how truth claims are always bound up with power relations (Colwell 1997). Bellicose history destabilizes what is taken for granted, and in doing so enables us to craft *alternative* histories of conquest, struggle, refusal and revolt that highlight the contingency of our current state of affairs and the possibility that things could be other than they are (Foucault 1998c).

Foucault's genealogies focused on a particular historical and geographical situation – the period between the mid-1700s and mid-1800s when European life experienced massive transformation. Struggles between, on one hand, the entrenched aristocracy and the remnants of feudal society and, on the other, the emerging bourgeoisie formed the backdrop against which a variety of seemingly disconnected developments occurred in military practice, medicine, psychological confinement and treatment, and penal practice. Foucault's archival work showed in minuscule detail how these developments were connected with the emergence of liberal political economic and moral philosophy. For example, transformations in French medical practice, such as how doctors visualized patients' bodies and the relation between these bodies and disease, corresponded with the ongoing re-configurations of social and political life around the promotion and protection of liberal individualism (Foucault 1994). From the viewpoint of total history, these transformations are part of the progressive development of scientific knowledge (on medicine and disease, and on the human body, both individual and collective) and political freedom (of sovereign subjects, both individual and collective), but Foucault shows how these apparently objective truths emerged through vicious conflict over how to understand, organize and govern life processes. His genealogies thus unsettle the sedimented foundations of Western metaphysics, modernism's linear and progressive narrative of History, and other sacrosanct institutions of modern European order, and show how these "truths" are instead contingent accomplishments that rest on nothing more or less than the play of force relations that could, at any time, reverse themselves.

Resilience is enmeshed in a similar series of conflicts and struggles over what forms the "truth" of complex life can assume, and how this complexity can be organized and governed. Indeed, Mark Duffield (2011: 763) characterizes resilience as "official response to environmental terror embedded in the radically interconnected and emergent lifeworld that liberalism has created." Duffield's use of the term "environmental terror" is crucial. He does not deny that contemporary liberal societies are characterized by complex interconnections and dependencies that enhance insecurity. But rather than asserting this as an ontological characteristic of life itself, Duffield's genealogical sensibility directs attention to trajectories of liberal governance that have actively *produced* these novel experiences of uncertainty and insecurity. He locates the emergence of environmental uncertainty in military and political developments during the twentieth century. In terms of military developments, the advent of gas and aerial warfare during World War I ushered in novel experiences of the environment as a potential source of threat. Without warning, an enemy attack could materialize that transformed habitable spaces of life into inhabitable spaces of death (Sloterdijk 2009; see also Chapter 5). Moreover, they also helped usher in new understandings of life as complex and uncertain. Air warfare introduced a new kind of military strategy that envisioned the enemy's production systems as interconnected networks. Strategic bombing targeted vital nodes in these networks. The logic here was that disrupting one part of the network would lead to cascading system failure, which would in turn lead to social disorder and erode the enemy's morale (Anderson 2010b). As Stephen Collier and Andrew Lakoff have detailed, civil defense and emergency preparedness activities developed in the United States during the 1950s as war planners reflexively turned this strategic military vision to their own vital infrastructure systems (Collier and Lakoff 2008; Lakoff 2007). The conceptual innovations military and civilian planners developed in response to these and similar challenges laid the groundwork for complex systems science, which, as Chapter 5 will detail, formed the foundations for ecological understandings of resilience.

Duffield and Collier and Lakoff each show how visions of life as complex and inherently vulnerable to disruption emerged not only out of developments in biology and economics, as detailed above, but also out of new developments in military planning. As the "official response" to neoliberalism's manufactured uncertainties, resilience is simply the latest fold in liberalism's struggle to govern social and ecological life. New techniques and strategies emerge in response to life's continued refusal to be governed as such – whether this refusal takes the form of self-organizing biological evolution, unpredictable market forces, worker organization and general strikes, or surprise air attacks.

Couched in its wider historical and political context, the emergence of resilience speaks to power's *inability* and repeated *failure* to govern life. Resilience does not reveal a novel truth about the nature of the world; instead, its renderings of indeterminacy emerge out of and respond to wider problems of governing recalcitrant social and ecological processes. As Chandler (2014a) astutely demonstrates, resilience gained traction as it resonated with neoliberal critiques of centralized economic governance and post-1968 debates on the left over how to govern and not be governed. Resilience provides a conceptual framework for understanding these refusals as expressions of life's complex interconnections and emergent temporality. It has proven particularly effective at rendering complex life governable – not because it promises to control recalcitrant social and ecological processes, but because it promises to develop individual and systemic capacities to adapt to change in ways that do not threaten established norms and institutions. In Brad Evans and Julian Reid's (2014) provocative phrasing, resilience initiatives create subjects who "forget how to die." They characterize adaptive capacities resilience instills as a kind of "non-death:" if, "to philosophize is to learn how to die," then "resilience cheats us of this affirmative task of learning how to die" (ibid.: 13). Here, death does not simply refer to the absence of biological life, but rather to a broader political-philosophical sense of becoming-otherwise. It signals an individual or collective's potential to become other than it is – to change, to take on different form, identity, function or content. The European Enlightenment's "death of God" – the dissolution of transcendentally-determined meaning and value – made death a part of life: because salvation was no longer guaranteed, life could become radically otherwise to itself. The absence of biological life was no longer one step on a divinely ordained path to salvation and eternal life, but now signaled life crossing a threshold to the great unknown. Rather than the continuation of life, death marked the limit of human knowledge and experience (Dillon 1996). Finite life – life that carries with it the possibility of death – thus embodies the potential to become radically other to itself (Malabou 2012). The various techniques of modernist and liberal security are efforts to forestall finite life's indeterminate potentiality. Resilience extends rather than displaces this biopolitical imperative (see Chapter 6): it provides liberal rule with a variety of concepts and techniques that enable it to visualize indeterminacy as complex life and manage complexity in ways that do not allow this radical becoming-otherwise to occur.

At first glance, resilience as a form of "non-death" might appear far removed from the concerns of ecologists and other proponents of resilience. But this "forgetting how to die" lies at the heart of ecologists' definition of resilience – and almost the same language. The ecologists Brian Walker and David Salt (2012: 3) describe this forgetting in terms of identity when they argue that resilience expresses how, "people, societies, ecosystems, and socio-ecological systems can all … be subjected to disturbance and cope, without changing their 'identity' – without becoming something else." The adaptive process of changing form and function without changing identity – "without becoming something else" – is precisely the kind of non-death that Evans and Reid critique. In their reading, the "non-death" resilience promises amounts to little more than surviving impending catastrophe. It does not seek to transform social and environmental realities to confront and remove the sources of vulnerability and insecurity, but rather allows life to live a little bit longer while the world crumbles around it. And this promise of extended life prevents the emergence of a more experimental ethic: as they ask in relation to the Anthropocene's catastrophic imaginary, why not experiment with a world about to end?

CONCLUSIONS

This chapter has explored some debates over resilience in geography and related fields. Throughout, I have tried to do so in a way that intentionally avoids attaching specific meaning to the concept. Instead, I have tried to maintain the *essentially contested* nature of resilience: it has a variety of competing definitions that each carry different and, in some cases, diametrically opposed ethical and political commitments. This contestation extends far beyond the disciplinary differences many scholars recognize. It is not a matter of different definitions in ecology, psychology, engineering, urban studies, disaster management, geography, and so forth. Instead, what is at stake with resilience is nothing less than the possibilities for understanding, living within, and managing indeterminate social and ecological relations. As genealogies of resilience have demonstrated, resilience is one particular way of intervening in this indeterminacy. A genealogical sensibility allows us to situate the essentially contested nature of resilience in political-philosophical terms. Its focus on the spatializations of meaning positions resilience as an outgrowth of conceptual and theoretical innovations, since the early twentieth century, that have grappled with the problem of how to understand and manage complex life that confounds modernist and liberal security techniques. Its focus on the beatified spaces of meaning foregrounds resilience as a site of ethical and political potential, a juxtaposition of non-stratified force relations saturated with the potential to create novel social and ecological arrangements out of the problems complex life poses to liberal order. And the bellicose history it offers situates this potential within struggles over how to orient thought and action in an indeterminate world.

Taken together, these genealogies suggest that the emergence and rapid adoption of resilience across a variety of fields of practice is more contingent – and hence malleable – than many proponents and critics acknowledge. Recognizing this bellicose history of resilience directs our attention to the possibility of *other* ways of understanding indeterminacy: other ways of knowing and relating to complexity, and other ways of knowing and relating to an uncertain and indeterminate future. It raises a different set of questions – not simply resilience of what and for whom, but how "resilience" as such has become the solution to surviving a turbulent future. And if resilience is one way of knowing and governing an indeterminate world, what are other ways? How have these other ways been vanquished, rendered scientifically or pragmatically inappropriate? This chapter has indicated some of these alternatives, such as political economic readings of vulnerability and resourcefulness, or post-colonial visions of resilience as working within and against overbearing pressures and everyday violence. The remainder of this book draws on the genealogical sensibilities outlined here to explore how resilience, as one (partial) solution to the problem of indeterminacy amongst others, shapes the possibilities for understanding and managing indeterminate social and ecological life. The next chapter examines how early formulations of ecological resilience offered a kind of subjugated knowledge on social and environmental surprise that prevailing positivist scientific and management approaches had marginalized.

FURTHER READING – GENEALOGIES OF RESILIENCE

The definitive critical engagements with resilience are David Chandler's *Resilience: The Art of Governing Complexity* and Brad Evans and Julian Reid's *Resilient Life: The Art of Living Dangerously*. Both offer detailed genealogies of the concept that locate its roots in post-1968 leftist critiques of centralized government and the emergence of complexity approaches in the life sciences, respectively. Claudia Aradau and Rens van Munster's *Politics of Catastrophe: Genealogies of the Unknown* situates resilience within a much wider genealogy of liberal strategies for governing uncertainty. Excellent introductions to Foucault's thought and genealogical style can be found in Charles Lemert and Garth Gillan's *Michel Foucault: Social Theory and Transgression*, Stuart Elden's *Foucault's Last Decade*, as well as selected chapters in *Space, Knowledge and Power*, edited by Jeremy Crampton and Stuart Elden (particularly Chris Philo's chapter on Foucault's *Society Must Be Defended* lectures). Gilles Deleuze characteristically puts his own spin on Foucault's *oeuvre* in *Foucault*, which helps distill many of the subtle maneuvers and critical tactics Foucault deployed throughout his corpus.

NOTE

1 With the term form of life, I follow the anthropologist Veena Das (2007), who draws on both Wittgenstein and Agamben (1978) to recognize forms of life as distinct conjunctures between language and practice. Das succinctly conveys their sense of form of life when she notes that infants have no prior knowledge of what the term "love" means; the meaning of love arises as the term is used in conjunction with certain affectionate actions. In this sense, the meaning of resilience likewise arises out of overlap between uses of the term and specific actions. Resilience "is" the practice of erecting barricades and installing CCTV security cameras around Manchester city center (see Box 2.1); it "is" field officers conducting transect walks through the foothills of Jamaica's Blue Mountains and it "is" those officers later sitting down with community activists to create a community disaster plan (see Box 2.2); it "is" the President of the United States exhorting citizens to return to malls and engage in mass consumption following the events of 11 September 2001. Each of these practices conveys something about the actors' ethical and political comportment towards other humans and non-humans as well as processes of change – in other words, they offer a window on a distinct form of life.

3

RESILIENCE AS SUBJUGATED KNOWLEDGE

INTRODUCTION

In the mid-2000s, the anthropologist Paul Nadasdy (2007) developed one of the earliest and still most incisive critiques of ecological resilience initiatives. His ethnographic research on the Ruby Range Steering Committee, a co-management initiative designed to promote wildlife conservation in Canada's Northwest Territory, conclusively detailed how adaptive co-management reinforced rather than challenged existing inequalities between indigenous First Nations peoples, on the one hand, and biologists, state officials, and private sector actors, on the other (Nadasdy 2003). These arguments challenged resilience scholars' claims that more participatory approaches to conservation and sustainable development would empower marginalized peoples. Rather than a simple technical project of folding in different knowledge claims (more on this in Chapters 4 and 6), Nadasdy showed that resilience initiatives are another stage in the ongoing drama of struggle over socio-ecological inequality and injustice. But his arguments did not stop there. He also took on resilience scholars' longstanding claims that the emergence of resilience exemplified a kind of Kuhnian paradigm shift in ecology. Following on from Thomas Kuhn's (1962) account of paradigm shifts as a response to phenomena that prevailing practices of scientific research and worldviews cannot account for, many proponents of resilience and the "new ecology" more broadly (Botkin 1990) argue that ecological research on resilience developed in response to scientific and management problems of complexity that confounded prevailing approaches centered on positivism and command-and-control management styles. However, Nadasdy, drawing on historians of ecology and environmental sciences (e.g., Hagan 1992; Demeritt 1994) showed that ecologists have long had an interest in otherwise contemporary problems, such as spatial interconnection, relationality and temporal emergence. In his reading, ecological work on resilience is less a paradigm shift than a shift in emphasis. This means that rather than disclosing a previously inaccessible truth about the nature of complex social and ecological systems, ecological work on resilience is one possible form of knowledge among others vying for influence and credibility. Resilience does not hold the keys to a new, more just and sustainable future in a complex world; instead, it is bound up in struggles over how to know and manage social and ecological life.

This chapter and the next push Nadasdy's arguments in a genealogical direction. As the previous chapter detailed, genealogy draws attention to the play of force relations – struggles, conflicts, strategies, and maneuvers – that structure the possibilities for knowing and acting in the world. This is precisely the kind of struggle in which Nadasdy situates resilience scholarship. But where he is quick to turn attention to the outcomes of this struggle – specifically, the growing influence of research on social and ecological resilience that fails to account for "the political economy of resource extraction that drives [resource] management in the first place" (Nadasdy 2007: 216) – I want to dwell on the elements of struggle. If we take the point that ecological resilience is a shift in emphasis on long-standing ecological problems of interconnection and emergence, the genealogical question becomes identifying what this shift involved and how the shift came about. What are the ontological and epistemological claims resilience scholars advanced, and what are the critical and strategic maneuvers these scholars deployed to create a space where these claims could have purchase on social and environmental management practice? While Chapter 4 details resilience proponents' strategies and practices of critique that denaturalized the claims of positivist science and command-and-control management practices, this chapter examines the particular understanding of social and ecological change resilience scholars developed. Starting with C.S. Holling's influential work on "ecological resilience" in the 1970s, I examine how ecologists developed a conceptual apparatus for understanding change structured around a topological mode of thought. As I will detail below, *topology* signals a way of thinking spatio-temporal processes that is not premised on absolute territorial division and quantitatively determined measurement (topography), but rather qualitative relations that can persist even as they undergo change and transformation. Importantly, a topological sensibility is attuned to processes of change and becoming rather than being and permanence. The topological sensibilities of ecological resilience enabled ecologists to advance a radically different understanding of knowledge, and the use of knowledge in social and ecological management initiatives, than we find in other understandings of resilience. This topological understanding of resilience is arguably the driving force behind many of the more counter-intuitive adoptions of resilience in a variety of professional fields (Chandler 2014a). However, set in the wider context of prevailing forms of thought on environmental research and management, this topological sensibility was marginalized in both academia and professional practice. In genealogical terms, resilience proponents' topological approach was a form of subjugated knowledge – an experience and account of environmental change that circulated at the margins of acceptable scientific and management discourse.[1]

I thus unpack the struggle Nadasdy (2010) describes as the conflictual and critical movement of subjugated knowledge from the margins to the center. This chapter maps the contours of this form of subjugated knowledge. The second section contrasts the topographical mode of analysis that typifies engineering resilience from the topological sensibilities that inform ecological resilience. The third section explores how this topological thought was formalized into ecological theories of the adaptive cycle (Holling 1986) and panarchy (Holling 2001; Gunderson and Holling 2002). The concept of panarchy provided ecologists with a unified and coherent way to account for social and ecological change that dramatically contrasted with prevailing understandings of sustainable development and environmental management. It thus became an important foundation for ecologists' ongoing critique of command-and-control management practices.

TOPOLOGIES OF RESILIENCE

Ecological understandings of resilience emerged during the 1970s and 1980s through a series of studies on predation, environmental degradation and environmental management conducted by C.S. Holling and colleagues (Holling 1973, 1978; see also Folke 2006). This work found that population levels of predators and prey did not follow the abstract laws of carrying capacity, which premised environmental management on identifying the optimal balance between predators and prey, and managing populations to sustain this balance. Instead, populations fluctuated in erratic patterns, displaying multiple equilibrium points: at some points in time, there would be high population levels of both predators and prey; at other times, the population levels of one species would decline and not return as carrying capacity calculations predicted. At each equilibrium point the ecosystems under observation maintained their basic structure and function. Holling's recognition that ecosystems could exist in a number of distinct steady states opened up a new field of inquiry: explaining and understanding the autonomous emergence of patterns of aggregation within an ecosystem (Levin 1998; Folke 2006). It also underpinned a novel understanding of resilience as the capacity of a system to autonomously shift between multiple equilibrium points in response to external disturbances. A resilient system would not simply return to a single optimal operating point, but might adjust and adapt to a new equilibrium where it continues to function as a viable ecosystem.

Holling's studies on predation were a groundbreaking intervention into orthodox thinking on ecological change at the time. Its implications were significant: Holling proposed nothing short of a radical rethinking of what it means to think about, study and manage the environment. The radical nature of Holling's work becomes clear when we look more closely at the differences between conventional (at the time) "engineering" approaches to resilience and the "ecological" understanding Holling and colleagues developed. Each of these is considered in turn.

The State Science of Engineering Resilience

In many of his writings, but especially following his groundbreaking 1973 paper, Holling carefully distinguished his understanding of *ecological resilience*, which emphasizes adaptability and topological transformation, from what he called *engineering resilience*, which focuses on resistance to disturbance and recovery from disruption (Holling 1986, 1992, 1996). The distinction gets to the core difference between the distinct disciplinary debates the previous chapter discussed. In Holling's reading, engineering resilience offers a reductionist view of ecological systems that assumes they are inherently stable and possess a single optimal operating point (Holling 1996; Folke 2006). It also assumes that a system is self-contained – that is, the system in question is hermetically sealed off from its outside, and its performance is not determined by its wider environment. While David Chandler (2014a) emphasizes that this understanding of simple complexity assumes there is an external vantage point from which the system can be controlled, Holling is concerned with the epistemological ramifications that follow. In a closed system, the number of variables that influence a system's behavior are necessarily limited and determinate. Because there is a single optimization point and a limited number of variables, researchers can precisely and accurately model a system's behavior. Deductive mathematical modeling allows them to reduce a system to a single independent variable whose behavior influences the behavior of other dependent variables within the system, and determine the system's optimal functioning level. Once researchers have confidently established the quantitative relationship between independent and dependent variables, they can more or less accurately predict how the system will behave as independent variables change. In theory, this predictive knowledge enables control: because the outcome of any intervention will be known in advance, the challenge is to determine what intervention will have the optimal outcome.

For Holling and colleagues writing during these formative years, this assumption that humans could fully know and control ecosystems was a major under-acknowledged source of contemporary environmental problems (Walker and Salt 2012). They developed an incisive critique against the "command-and-control" management strategies that resulted from visions of engineering resilience. In important language, Holling and Gary Meffe define the logic of command and control as an

> approach to solving problems ... in which a problem is perceived and a solution for its control is developed and implemented. The expectation is that the solution is direct, appropriate, feasible, and effective over most relevant spatial and temporal scales. Most of all, command and control is expected to solve the problem either through control of the processes that lead to the problem (e.g., good hygiene to prevent disease, or laws that direct human behavior) or through amelioration of the problem after it occurs (e.g., pharmaceuticals to kill disease organisms, or prisons or other punishment of lawbreakers). The command-and-control approach implicitly assumes that

> the problem is well-bounded, clearly defined, relatively simple, and generally linear with respect to cause and effect.
>
> (Holling and Meffe 1996: 328)

Their definition points to the intimate connection between engineering resilience and command and control: both share an assumption that systems are inherently stable and possess a single, optimal operating point that can be identified through statistical analysis. Command and control is a *general approach to problem-solving* that can be applied in a number of fields – public health, jurisprudence, medicine, and the criminal justice system. In each case, it involves a certain relation between managers and the "object" they are attempting to manage, whether this is human communities, biological populations or ecosystems: there is a privileged "center" that determines appropriate interventions through projecting the impacts of different interventions and selecting the optimal outcome. In the case of ecology and environmental management, this could involve identifying maximum sustained yields for resources, such as the number of fish that fisherfolk can harvest or the number of trees that loggers can fell while ensuring future resources will be available. It could also involve calculating the carrying capacity of an ecosystem and devising and carrying out environmental management strategies that will prevent plant and animal population levels from exceeding this safe limit. In each case, command and control relies on hierarchy. This is so in two senses. First, scientists and government planners claim expertise and authority over the objects of management, because they alone have access to objective and totalizing visions of how the social and/or ecological system in question operates. Second, hierarchy also expresses what we might think of as a *semiotics* of optimization (cf. Guattari 1996). In brief, the assumption that optimization is feasible and desirable imposes a semiotic hierarchy that constricts the possible meanings scientists, planners and the public might give to social and environmental phenomena and processes: they are apprehended in terms of their deviation from optimal conditions, and subject to normalizing correction if they are suboptimal. And of course, only scientists and planners possess the knowledge and skill to identify this optimization point and manage the system so that it sustains optimal functioning.

As well as a general technique of problem-solving, Holling and Meffe's (1996: 328) insistence that "the solution is direct, appropriate, feasible, and effective over most relevant spatial and temporal scales" indicates that command and control carries implicit assumptions about the spatiality of problems and, by extension, "life" in general. This is a vision of space that is bounded and objectified – what geographer David Harvey (1973) refers to as *abstract space*. Elsewhere, Holling (1996: 34) asserts that engineering resilience is underpinned by abstract conceptualizations of space that enable it to, first, reduce the complexity of social-ecological systems to a single, discrete, and measurable variable (for example, carrying capacity defined by population densities of predator and prey), and second, to discover the underlying order of the system that reveals its optimal performance.

This is an important point that deserves unpacking, for his argument helps us recognize how engineering resilience is premised on a specific mode of thought, or way of reflecting on and making truth claims about the world, that ecological resilience challenges. Specifically, engineering resilience emerges through a *calculative mode of thought* firmly entrenched within modern metaphysics. Calculation here is more than quantification, counting, adding and division; it involves a general science of order, what Michel Foucault (1989) calls a *mathesis* in which all relations between beings are conceived in the form of measure and order. The political geographer Stuart Elden (2001, 2006) has extensively detailed how the modern *mathesis* is underpinned by a Cartesian vision of space as an abstract, infinitely divisible, extensive, and measurable surface. Elden's careful unpacking of the modern *mathesis* helps us recognize the stakes of Holling's critique of engineering resilience. In brief, Elden's (2006) genealogy of territory shows that this abstract conceptualization of space does *not* extend back to ancient Greeks – and particularly Euclidean conceptualizations of space – but instead turns on René Descartes' radical reading of geometry, which continues to hold both metaphysical and political purchase today. For Elden, the mathematical grounding of Western metaphysics – the *mathesis* of engineering resilience Holling targets for critique – emerges only in the seventeenth century with Descartes' attempt to position extensive space as the essence of corporeality. For Descartes, the surface of any body – defined by the three dimensions of "space" – remains constant even as motion and position change. All being comes to be seen as calculable: all that exists is essentially objective surfaces that can be divided into discrete units, ordered, arranged, measured, quantified – in short, subject to *mathesis*. This is the heart of Descartes' geometry: it turns geometry, which was once a mode of thinking continuum and connection of the immeasurable, into an algebraic method for dividing, measuring, and analyzing surfaces as discrete units (Elden 2001, 2006). Engineering resilience and its underlying Cartesian geometry are forms of what Deleuze and Guattari (1987) call "State Science," or abstract and objectifying modes of thought that enable the world to be increasingly brought under the sway of a calculative and controlling rationality. The essence of all beings becomes perfectly knowable through a geometry that quantifies, measures and analyzes the dimensions of "space" that define a surface (Elden 2006). Cartesian geometry makes finite and total knowledge of the world possible for the first time; this knowledge enables prediction, analysis, mastery, and control of all beings grasped as essentially calculable (Foucault 1989).

The Nomad Science of Ecological Resilience

While not couched in terms of Deleuzo-Guattarian philosophy or Foucauldian genealogy, Holling's pathbreaking 1973 paper on "Resilience and Stability of Ecological Systems" nonetheless posed a direct challenge to the State Science of engineering resilience. He begins the paper with a consideration of different theories and modeling methodologies for the study of predator–prey relationships. Despite different starting assumptions, each tends to provide a consistent view of ecosystem change: either the system will tend towards global instability or global stability, and that stability tends towards a stable limit cycle, or constant cycle of fluctuations between population levels of predators and prey. Of course, this is a variant of engineering resilience, in which resilience is simply the capacity of the system to return to equilibrium after a disturbance. In contrast, Holling uses examples from both closed and open ecosystems to argue that even closed systems display multiple, distinct domains of attraction. We will detail this term below, but for now, *multiple domains of attraction* essentially means that systems have multiple equilibrium points. One example Holling cites involves a closed freshwater lake system: an experimental fishery that sought to improve salmonid stocks in Lake Windermere by reducing the numbers of perch (a competitor species) and pike (a predator species). The perch population fell dramatically during the intervention, but in the two decades that followed, it failed to return to previous levels. For Holling, this indicates that there are multiple stable configurations of predator–prey relationships within the same closed system. This simple insight has profound implications. First, it directs analytical attention to *processes* and *relations* such as fecundity, predation and competition that cause observed population behaviors. At issue here is not the quantitatively determined level of each population, but rather, in Holling's (1973: 9) words, the "processes that link organisms." Second, Holling argues that multiple equilibrium systems display resilience *rather than* stability. Here, Holling (ibid.: 14) defines resilience as "a measure of the persistence of systems and of their ability to absorb change and disturbance and still maintain the same relationships between populations or state variables." Resilience is the *converse* of stability: citing the example of periodic budworm outbreaks in eastern Canada, Holling argues that the budworm community is highly unstable, but that this instability gives it resilience: it has the capacity to absorb shocks, reorganize itself, adapt to unpredictable environmental conditions, and "persist" through changing.

This definition is one of the first detailed specifications of "ecological resilience." As both David Chandler (2014a) and Brad Evans and Julian Reid (2014) have argued in terms of political philosophy, Holling's definition *transvalues* instability and insecurity. This means that ecological resilience, for Holling, transforms the meaning, significance and value of instability. If engineering resilience saw instability as a condition to be avoided, and defined resilience as the antithesis of instability, Holling's definition of ecological resilience asserts that resilience emerges *out of* instability. Instability *can* prove beneficial to systems, as it gives them the flexibility to reorganize themselves in response to disturbances. Indeed, Holling is quite explicit about the challenge his vision of resilience poses to conventional ways of understanding ecosystemic change. He argues that:

> instability, in the sense of large fluctuations, may introduce a resilience and a capacity to persist. It points out the *very different view of the world that can be obtained* if we concentrate on the boundaries to the domain of attraction rather than on equilibrium states. Although the equilibrium-centered view is analytically more tractable, *it does not always provide a realistic understanding of the systems' behavior.*
>
> (Holling 1973: 15, emphasis added)

The italicized sections are key: in the former, Holling advances an epistemological claim: focusing on "the boundaries to the domains of attraction" instead of "equilibrium states" provides a different way of knowing the world. In the latter, Holling presents an ontological critique: while the "equilibrium-centered view" of engineering resilience allows for elegant analysis, it fails to capture the reality of system behavior, and specifically, the dynamic movements within and across multiple equilibrium points. Indeed, "ecological" definitions of resilience reject the Cartesian reduction of ontology to measure. Ecological resilience refuses to take a stable, pre-constituted, extensively and quantitatively defined system or individual body as its starting point. Instead, it sees any stability as a transitory outcome of a more fundamental interplay of dynamic relations and processes between a few key elements that drive a system's behavior.

Holling's understanding of ecological resilience opened a new line of questioning into the *processes* of self-organization: how do systems autonomously adapt to, and reconfigure themselves, in response to exogenous change? This is a question that requires a distinct mode of thought: not the Cartesian *mathesis* of engineering resilience, but rather a *topological* sensibility that focuses on dynamic processes of change and transformation. In contrast to the absolute space of Cartesian algebra, *topology* is a term derived from differential geometry to signal a multi-dimensional *state space* that can be stretched, twisted and deformed while retaining the same essential structure (Martin and Secor 2013). State space is not defined by abstract, quantitatively determined measurements; instead, it consists of qualitative relations between the variables that constitute a particular system. More specifically, state space refers to all possible combinations and configurations of these variables (DeLanda 2002; Jones 2009). For example, we might say that the state space of a freshwater lake undergoing eutrophication consists of four key variables, or dimensions: land use in its watershed, phosphorus levels in the water, vegetation types surrounding the lake, and the amount of algae in the lake (after Gunderson 2000). Each of these variables will change at different speeds. Land use in the watershed may change slowly, on the timescale of decades, for example, as people take up farming, encourage forest regrowth, or increase suburban development. Phosphorus levels in the water may fluctuate more rapidly, over a timescale of months. It may be high during the months that farmers plant and fertilize their fields; and low during other times of the year. If the system is resilient – able to adapt to change – it will maintain the same basic levels of service provided despite these multiple layers of ongoing change: for example, lake water will remain clear.

This basic level of ecosystem service is a "basin of attraction," or a region in state space in which the system tends to remain as its variables change. A basin of attraction is anchored by an attractor, which functions much like a small weight dropped onto the fabric of topological

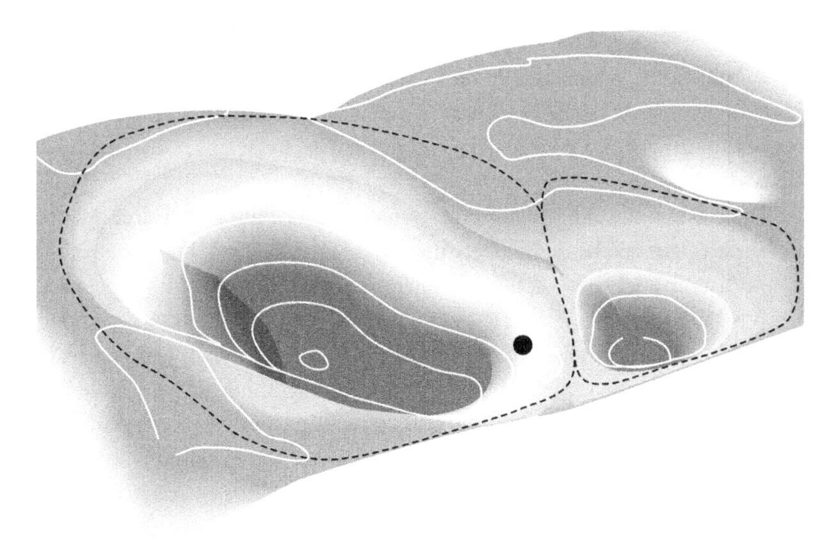

Figure 3.1 Stability landscape representing two basins of attraction (depicted by dark shading).The dotted line represents the threshold or transition between basins. The ball (or dot) represents the system in its current position. The system is resilient if it stays within that basin of attraction and does not cross a threshold: any changes will be topologically indistinct from one another. Once it crosses the threshold, then it loses its identity, form and function, and is no longer resilient.

Adapted from Figure 1b in Walker *et al.* (2004).

state space: an attractor creates a concave depression that pulls a system's trajectory towards it as the system moves through state space (see Figure 3.1). In our example of the lake, an attractor may be a certain quality of the water: clear, indicating a "healthy" lake, or turbid, indicating a "degraded" lake. Thus our state space has two basins of attraction: the lake can be clear and healthy, or cloudy and degraded. As the constituent variables of the system constantly change – as people engage in new activities, as plants and animals respond, and so forth – a resilient social-ecological system will likewise constantly reorganize itself in ways that absorb these changes and allow the lake to remain clear. In a resilient system, each reorganization is topologically indistinct from that which came before it – that is, the system continues to provide the same functions despite undergoing quantitative changes in its constituent variables. The ecosystem occupies a new point within state space that nevertheless remains in the same basin of attraction. Thus, there are multiple "states" in which the system can exist that enable it to maintain a certain quality or performance, not just a single equilibrium point that represents a perfectly healthy lake and deviations from this point that must be corrected.

The essence of such multi-stable systems lies not in discrete and quantifiable parts that can be deductively measured and analyzed, but rather in the *relative* distance between each non-metric variable that defines its state space. Indivisibility is a characteristic of topological visions of space that underpin ecological definitions of resilience: topology gestures towards a space that can be stretched, twisted, defigured, and deformed, all without losing its basic structure and function. In this sense, topology is a form undergoing constant (de)formation. What matters is not the quantitative relation between constituent variables, but rather the qualitative relations between them, and the persistence of these relations as the system is exposed to disturbance: as a system undergoes change, can it reconfigure itself (or in other words, occupy a new point in state space) in ways that will preserve the qualitative relations that give the system its identity and function? Here a simple, straightforward example will illustrate the point. In topological terms, a donut and a coffee mug are indistinct (see Figure 3.2). Stretch out

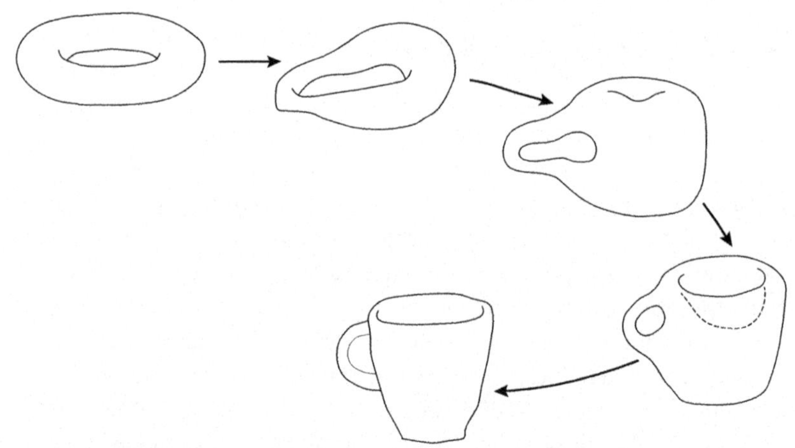

Figure 3.2 Topological equivalence of a donut and a coffee mug: stretching and de/reforming can turn a donut into a coffee mug (and back). Both represent a solid volume with a single hole.

Figure adapted from Stephen Barr (1964), *Experiments in Topology*. Mineola, NY: Dover Publications.

one side of a donut (assuming it is pliable), make an indentation in the top: a coffee mug. While these objects are quantitatively distinct – they have unique measurable dimensions of height, width and length – they are *topologically indistinct*. Both are three-dimensional surfaces with a single hole. So long as the stretching or indenting does not tear a new hole or collapse the existing hole, the qualitative relation (a surface with a single hole) persists even as the object changes form, function and identity.

Rather than a metric space striated by limits, borders, boundaries, and other techniques of measure and division, the smooth space of topology suggests an ontologically prior connectivity and relationality (Deleuze and Guattari 1987: 361; DeLanda 2002). Ecological resilience is thus attuned to dynamic forces and interrelations that precede the appearance of any form of measure or stability; it operates through the differential geometry that precedes Cartesian geometry (Holling 1986, 1996). It signals not an extensive quantity that can be measured and analyzed for its deeper order, but an intensive quality that reveals nothing more than the immanent surface relations between its constituent elements.

Thus, in focusing on processes, ecological resilience refers to the dynamics of topological equivalence: how ecological systems' movements in state space either enable the system to remain in the same domain of attraction or move into a new one. A system's ability to absorb change and reorganize while retaining its basic structure is determined by its relative distance to the threshold that marks the limit between one domain of attraction and another, and the landscape or topography of this distance. Ecologist Brian Walker and colleagues provide four defining aspects of resilience in these terms (Walker *et al.* 2004). First is a system's latitude, or the amount of change it can undergo before reaching a threshold. This is essentially the 'diameter' of the basin, or the area of state space in which the system can move. Second is the system's resistance, which refers to the ease of change within a system. This is roughly equivalent to the 'height' of the basin, the relative distance between nearby points in state space that must be traversed in order to effect change. Third is the precariousness of a system. This refers to the system's position in state space, or its distance from a threshold. Fourth is *panarchy*, a neologism developed by Holling (2001; see also Gunderson and Holling 2002) that refers to the multi-scalar dynamics that drive change and adaptation within a system. I will return to this concept below, for panarchy signals a decisive break in how ecologists conceptualize ecological resilience. For now, the point is that these defining characteristics of ecological resilience all concern, in one way or another, the *limits* of a system and its *potential* to transgress these limits. This is precisely Holling's argument, in a quote referenced above, when he urges readers to consider, "the very different view of the world that can be obtained *if we concentrate on the boundaries to the domain of attraction rather than on equilibrium states*" (Holling 1973: 15, emphasis added). In short, ecological resilience is concerned with the potential for a system to become other than it is – whether this becoming is a process of adapting to change through reorganizing into a topologically equivalent form or maladaptation that responds to change in ways that lose form and function.

Ecological resilience thus confronts the State science of engineering resilience with the *nomad science* of morphogenesis, or the emergence of objective and individuated beings out of pre-individual forces (Deleuze and Guattari 1987; Simondon 1992; DeLanda 2002; cf. Holling 1992). It emerges through a topological mode of thought attuned to flow, change, and becoming rather than presence, centeredness, and being. Against assumptions of homeostasis that frame change as a breakdown or deviation from a perfect or optimal norm, and thus something to be avoided, ecological resilience presents a vision of change as necessary and beneficial – if properly managed. Against assumptions that perfect knowledge and prediction are both possible and desirable, ecological resilience offers a non-linear account of change formed through random events that push socio-ecological systems across thresholds. Because it strikes a potentially fatal blow against positivist and empiricist assumptions that underpin State science, it is often interpreted as a radical alternative that introduces critical modes of thought into environmental governance (e.g., Sneddon 2000; Boyd and Juhola 2009; Yamane 2009). However, what interests me here is not the absolute distinction between these forms of science but rather their interconnection, for there can be no such thing as pure "State" and "nomad" science (Deleuze and Guattari 1987). Following Deleuze and Guattari, my interest lies in how State science appropriates and imposes its order on the inventions of nomad science. This is a question of how the topological mode of thinking that characterizes ecological resilience becomes enfolded into governmental programs of social and environmental management. Accordingly, the next two sections explore, first, how ecologists have developed and advanced the notion of ecological resilience on the back of Holling's early work, and second, how the topological thought of ecological resilience is transforming the way researchers and practitioners understand and manage societies and environments as complex social and ecological systems.

ACHIEVING SUSTAINABILITY THROUGH RESILIENCE

Holling's 1973 paper signaled a decisive break with mechanistic understandings of ecosystems and a linear understanding of change and causality. However, while this work is now rightfully recognized as a field-defining publication, its wider interdisciplinary influence was somewhat muted for many years immediately after its publication, in no small part because it went against the grain of dominant environmental thought at the time. Holling's analysis was published on the heels of a number of signal events in the modern environmentalist movement: the 1972 United Nations Conference on the Human Environment, held in Stockholm, Sweden, The Club of Rome's 1972 publication of its influential *Limits to Growth* thesis, and James Lovelock's 1972 publication of his influential "Gaia thesis." These publications were considerably more in tune with prevailing understandings of the global environment and environmental problems than Holling's work on multiple-equilibrium systems. Lovelock's "Gaia thesis" presented, in a punchy and immediately graspable term, a vision of the Earth as a single complex system that maintains biological and ecological life on the planet. In turn, The Club of Rome's *Limits to Growth* asserted that exponential population growth would soon surpass humanity's technological capacities to provide resources for its ballooning population. While there are significant parallels to Thomas Malthus' controversial theory of population (Robbins 2004), what is significant here is how *Limits to Growth* essentially set up a problem of optimizing ecosystem carrying capacity: it visualized population and resources in a linear relation with a single optimal point of maximum efficiency. Moving beyond this optimal point – i.e., outstripping the capacities of both Earth and human technology to provide resources for a growing population – would lead to a Malthusian nightmare of resource shortages, hunger, starvation, and suffering for the whole of humanity. This all-or-nothing imagery is closer in form and spirit to engineering resilience and its vision of systemic stability around a single optimal operating point than Holling's theory of multiple equilibrium systems and ecological resilience. And indeed, *Limits to Growth* was heavily influential at the 1972 Stockholm Conference, which asserted, among other things, the need to limit pollution, avoid exhausting non-renewable resources, ensure Earth's ability to produce renewable resources, and promote more environmentally sensitive economic development through better state planning and environmental management (Adams 2001). The latter point is key: the Stockholm Conference institutionalized a vision of environmental management premised on centralized, command-and-control-style interventions in order to ensure that human activity would not outstrip its limits to growth. In a sense, Holling's arguments were both out of time and out of place. They were certainly an outgrowth of the growing interest in complex systems theorizing across diverse fields such as military planning, economics, public administration, and computer science – something Chapter 5 will explore. But his topologically tinged slant on complexity had yet to take hold in environmental sciences, dominated as they were by visions of single-equilibrium systems, the problem of optimization, and command-and-control management strategies designed to achieve sustainable development.

Thus, during the period from 1973 to the publication of Holling's influential *panarchy* thesis (Holling 2001; Gunderson and Holling 2002), work on ecological resilience extended on two simultaneous fronts: on one hand, conceptual development focused on understanding the dynamics of multiple-equilibrium systems and exploring the scalar relationships between these dynamics and global changes; and on the other hand, a concerted critique of prevailing scientific knowledge and conventional management styles to create a space for the alternative view of the world ecological resilience offered. The next section explores the former; the latter is the focus of the next chapter.

FORMALIZING TOPOLOGY: FROM THE ADAPTIVE CYCLE TO PANARCHY

Given the focus on global environmental issues during the 1970s and 1980s, proponents of ecological resilience were particularly interested in understanding the relation between the dynamics of localized and bounded multiple-equilibrium systems and global change. This concern found its way into the title of Holling's important 1986 chapter, where he develops an understanding of how ecosystems change that relates local actions and events to global biospheric changes. The solution he arrives at is the preliminary formulation of what he would later call the *adaptive cycle*. This concept is a heuristic device for understanding how change occurs in multiple-equilibrium systems. As he elaborates in later essays (Holling 1995, 2001), the adaptive cycle recognizes that a system can be characterized by three properties at any point in time: wealth, or potential for change; controllability, or the degree of connectedness and rigidity within a system; and adaptive capacity, or the system's resilience. These properties will change over time, and will affect a system's level of organization and disorganization. In short, the adaptation cycle describes the movement of a system between periods of organization and disorganization, driven by the accumulation and destruction of potential for change. This movement occurs through four distinct phases (see Figure 3.3): exploitation (r), conservation (K), release (Ω), and reorganization (α). First is exploitation, also known as the r-phase, in which connectedness and wealth are low. This is the period in which structure and connectedness begin to form, and which provides the basis for accumulation of potential. Second is conservation, the K-phase, which is characterized by high wealth and high connectedness. In the K-phase, the system's wealth or potential will often be under-exploited, because it is bound up in rigid structures

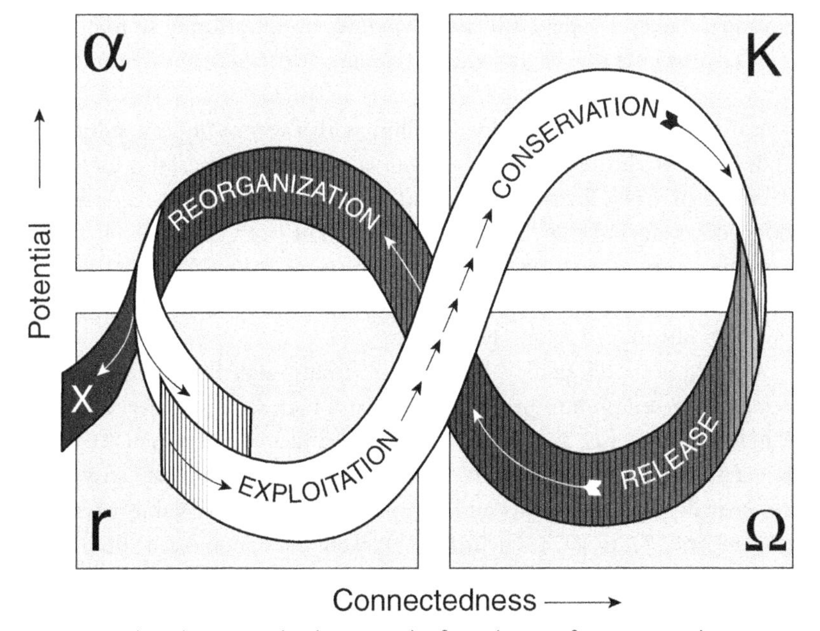

Figure 3.3 The adaptive cycle, depicting the four phases of ecosystem change. Figure from Figure 4 in Holling (2001).

that prevent its utilization. Third is release, the Ω-phase, a period of creative destruction that emerges when a rigid structure is destabilized. Connectedness here is high but wealth is low. Fourth is reorganization, the α-phase, which is characterized by low connectedness and high potential.

When Holling first sketched out the adaptive cycle, ecologists had long recognized the r–K movement within ecosystems: in essence, it describes the process of ecosystem succession as species migrate into a new region and exploit available resources (r) and establish themselves within that system (K). However, the back loop was a much more radical notion. Processes of release and reorganization signal moments of self-organized creative destruction where a system reorganizes itself in response to change.[2] In other words, the back loop refers to processes of topological transformation, in which a resilient system reorganizes itself in response to external pressures. Holling later emphasized how the back loop is the locus of topological plasticity:

> The reshuffling in the back loop of the cycle allows the possibility of new system configurations and opportunities utilizing the exotic and entirely novel entrants that had accumulated in earlier phases. The adaptive cycle opens transient windows of opportunity so that novel assortments can be generated. (Holling 2001: 397).

The adaptive cycle thus provided Holling with a way of understanding *how* systems move through phase space *without* crossing a critical threshold and losing their basic form, identity and structure. Topological reconfiguration occurs during the "back-loop" movement from Ω–α, which allows systems to autonomously self-organize in response to disturbances. The adaptive cycle thus describes the dynamics of ecological resilience – the problem introduced by his 1973 paper.

But that is not all. In his 1986 chapter, Holling also drew on established ecological research into patch dynamics and hierarchy in order to understand how systems interacted with social and environmental forces at larger and smaller scales. In brief, patch dynamics asserts that ecosystems are spatially heterogeneous and comprise many smaller sub-ecosystems (Pickett and White 1985; Levin 1992; Pickett and Cadenasso 1995). Hierarchy theory, in turn, recognizes that variables within an ecosystem tend to have high degrees of interaction amongst themselves, but limited interaction with processes at larger and smaller scales. *Hierarchy* here signals less a theory of hierarchical command, and more a way of understanding how variables within "semi-autonomous levels" of a sub-system interact with each other and with sub-systems at larger and smaller scales (Holling 2001: 392). Ecologists derived this theory of hierarchy from Herbert Simon's (1962) work, "The Architecture of Complexity," which introduced a number of important innovations for understanding complexity as a dynamic, multi-scalar and non-equilibrium process (Holling 1995). Chapter 5 will detail Simon's direct and indirect influence on ecological resilience; for now, we can see Simon's understanding of hierarchy in Holling's (1986: 77) assertion that scales represent ecosystems, which he defines in Simonian terms as, "communities of organisms in which internal interactions between the organisms determine behavior more than do external biological events." Blending this understanding of hierarchy with patch dynamics and multiple-equilibrium systems enabled Holling to re-conceptualize complex ecological system dynamics and the processes through which they change. Specifically, unexpected and non-linear changes result from the complex cross-scalar interactions. Hierarchical relations across scales are thus the source of surprise and uncertainty in complex ecosystems (Walker *et al.* 2006). And the process of self-organized adaptation results from nothing more or less than the non-linear interactions of key elements *within* sub-systems that are bounded by processes occurring at larger scales – even as these processes are themselves impacted through complex relations with other social, economic and biophysical systems.

The importance of this argument cannot be overstated. With this understanding of complex, hierarchical linkages across scales, Holling effectively developed the first specification of ecological resilience that linked together social and environmental systems (Walker and Cooper 2011). Ecologists could now lay claim to resilience as a way to describe the dynamics of social change, a contentious point that Holling and colleagues would develop further in later work (Holling 2001, 2004; Westley *et al.* 2002). On this point, they were merely echoing Herbert Simon (1962, 1996), who argued decades before that hierarchy described not only social and ecological dynamics but also psychology, genetics, linguistics, design thinking, and cybernetics. Nonetheless, Holling's (1986) intervention linked the complexity of social and ecological systems around the category of resilience.

During the 1990s, ecologists further refined Holling's understanding of global–local linkages into the formal theory of *panarchy* (Gunderson and Holling 2002; see also Holling 2001; Gunderson *et al.* 1995a). This

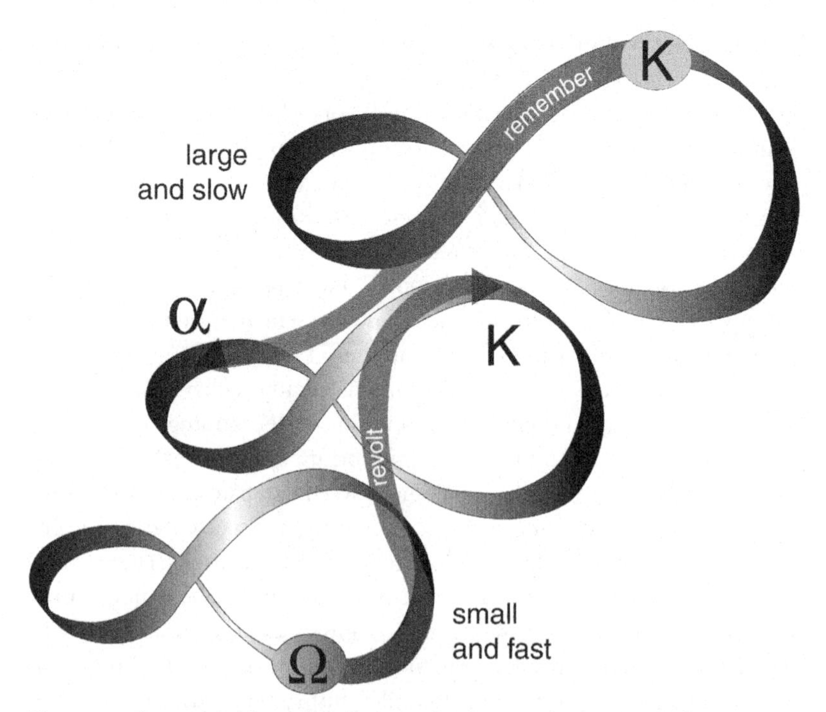

Figure 3.4 Panarchy: hierarchically organized, nested adaptive cycles.
Figure from Figure 1 in Garmestani and Benson (2013).

neologism essentially appropriates Simon's concept of hierarchy, and specifically his elaboration of complex systems comprising limited connections between semi-autonomous scales. However, ecologists substituted the prefix *pan-* to indicate Pan, the Greek god of nature, and also to differentiate Simon's dynamic understanding of hierarchy from a rigid, structure-bound sense that prevailed in ecology at the time (Holling 2001). Panarchy offers a vision of complex social-ecological systems as nested sets of adaptive cycles that correspond to sub-systems operating at particular organizational scales and speeds (Figure 3.4). For example, the time-space of an insect population within a small part of a forest may change over a matter of years; the trees in which they live have a larger time-space of many decades covering the entire forest; while the forest has a time-space of centuries and will be embedded within a larger ecosystem, and so forth (e.g., Folke 2006). The intersections of these slow- and fast-moving variables and the communication between their different levels are the key for understanding ecosystem dynamics. Slow-moving variables, those with a larger time-space, determine the dynamic behavior of a system: they set parameters on the change that is possible during periods of reorganization at faster, smaller scales (Walker *et al.* 2002). Slow-moving variables cascade down the panarchy from the K-phase of large-scale processes into the α-phase of reorganization at smaller-scale sub-systems, a process Holling (2001) calls "remember." This movement constricts the possibilities for change and generally ensures the stability of the system's overarching structure despite ongoing change at smaller and more specific levels. Change at higher levels of the panarchy is rarer than at faster and smaller levels; it involves a process of "upscaling" change from the Ω-phase of lower levels to the K-phase of slower, larger levels. This process of "revolt" (Holling 2001) can only occur if the larger sub-system is in its K-phase and is thus extremely rigid and susceptible to shock, in which case a fast variable's destabilizing force will knock a slower one off-balance. If the larger sub-system is not highly rigid and inflexible – that is, if it is resilient – changes in lower levels will have little if any effect, and the system will absorb these changes through adaptation.

Panarchy thus provided the solution to the problem of global change ecologists confronted in the 1970s. Not only did it riff on complexity science's topological thinking to link the social and the ecological, it also offered a theoretical framework for understanding how complex, multiple-equilibrium dynamics within more or less bounded ecosystems interacted with processes occurring at larger scales, which conceptually linked the adaptive cycle with global biophysical changes. In the process, panarchy also re-framed the problem of sustainability in terms of ecological resilience. Holling signaled this reframing when he described the relation between growth and conservation (r–K) and creative destruction and reorganization (Ω–α) in the following terms:

> It is as if two separate objectives are functioning, but in sequence. The first maximizes production and accumulation; the second maximizes invention and reassortment. The two objectives cannot be maximized simultaneously but only occur sequentially. And the success in achieving one inexorably sets the stage for its opposite. The adaptive cycle therefore embraces two opposites: growth and stability on the one hand, change and variety on the other.
>
> (Holling 2001: 395)

Understood in these terms – as "embracing" the opposites of "growth and stability" and "change and variety" – the adaptive cycle provided an inventive slant on the problem of sustainability. Its emphasis on self-organizing, topological transformations reconfigured sustainability from a linear problem of optimizing scarce resources into a non-linear problem of systemic adaptation to emergent social, economic and environmental conditions. Together, the concepts of panarchy and the adaptive cycle allowed ecologists to reposition environmentalist concerns with sustainable development within the "very different view of the world" (Holling 1973: 15) ecological resilience offered. Sustainability could now be framed as a matter of building resilience.

CONCLUSIONS

The publications of Gunderson and Holling's (2002) co-edited volume and Holling's (2001) theoretical elaboration of panarchy were watershed moments in ecological research on resilience (Holling 2004). With the benefit of hindsight, we can see that they brought the topological sensibilities of ecological resilience out of marginal corners of ecology into mainstream environmental research. Of course, this is not to say that resilience thinking supplanted other ways of studying human–environment relations. Instead, panarchy provided a foothold for ecologists to intervene, to an unprecedented extent, into ongoing debates over pressing social and ecological problems of sustainability. To be sure, by the early 2000s the landscape of environmental studies and environmental management had drastically shifted. Sustainability had become firmly ensconced as the guiding principle for social and ecological governance (Whitehead 2007; While *et al.* 2010), but the problems it addressed no longer revolved solely around avoiding future environmental disruptions. Instead, many effects of climate change were already becoming visible. Sustainability was no longer simply a problem of finding ways to mitigate the environmental impacts of continued economic growth and development; it now had to confront the challenge of how social and ecological systems might adapt to unprecedented changes. The impact of adaptation on environmental management cannot be overstated, for it points to a new understanding of the relationship between humans and the non-human environment. Early interest in sustainable development was premised on command-and-control logics, a form of simple complexity that maintained the possibility of a position *outside* the system in question. From this external vantage point, scientists and planners can obtain an objective and predictive view of the system, and design interventions that will bring about desired environmental outcomes. In contrast, the recognition that humans are already adapting to climate change erodes this boundary. Adaptation implies that humans are always already *within* a dynamic system, reacting to changes they cannot foresee. Here, sustainability is less a matter of designing interventions that will achieve the optimal balance between economic growth and environmental conservation than building-in adaptive capacities to identify and respond to potentially damaging environmental changes as they occur. Of course, this systemic capacity for autonomous self-(re)organization is precisely what ecologists had been studying through ecological resilience. Ecologists were well positioned to intervene in sustainability debates as adaptation increasingly came to the fore (see Pelling 2003, 2010). They had cleared a strategic position through decades of sustained critique against command-and-control approaches to environmental management, to which we now turn.

FURTHER READING

Brian Walker and David Salt's *Resilience Thinking: Sustaining Ecosystems and People in a Changing World* provides an accessible and thorough description of ecological resilience theory's core elements. They summarize more detailed theoretical discussions on topological thinking in ecology found in C.S. Holling's challenging 1992 *Ecological Monographs* essay, "Cross-scale morphology, geometry, and dynamics of ecosystems," as well as topological sensibilities peppered throughout Lance Gunderson and C.S. Holling's influential co-edited volume *Panarchy: Understanding Transformations in Human and Natural Systems*. Readers interested in topological thought as it is more broadly deployed in geography and related disciplines would do well to consult Lauren Martin and Anna Secor's 2013 *Progress in Human Geography* article, "Towards a post-mathematical topology," as well as Stephen Collier's 2009 *Theory, Culture & Society* essay on "Topologies of power: Foucault's analysis of political government beyond 'governmentality.' "

NOTES

1 On "subjugated knowledge" in policymaking, see Grove's (2013a) analysis of the subjugated knowledge of mitigation in Jamaican disaster management.

Much like resilience signaled an alternative experience of sustainability for certain ecologists that could not be slotted into prevailing environmental management frameworks organized around principles of centralized command and control, mitigation signaled an alternative experience of disaster that could not be slotted into prevailing disaster management frameworks organized around principles of preparedness.

2 Holling (1986) explicitly cites Austrian economist Joseph Schumpeter on this term; see Connor Cavanagh's (2016) detailed analysis for the epistemological and political implications of this conceptual borrowing.

4

RESILIENCE AS CRITIQUE

INTRODUCTION: CRITIQUE BEYOND CRITICAL THEORY

For many critical social scientists, critique is the exclusive province of radical academic and activist practice. Critique, as a practice of questioning and denaturalizing implied assumptions about the way the world works – ontological assumptions about what exists, epistemological assumptions about how we can know this world, and ethical assumptions about how we should act in this world – is a device for showing the *contingency* of what we take for granted: that things could always be other than they are. It contrasts with applied research, which seeks to produce knowledge that can be useful for policymakers, corporations, and other institutions with a vested interest in the status quo. Critical scholars thus tend to assume that critique will destabilize claims to expertise in ways that undermine the status quo and generate possibilities for transgression – that is, for inventing new ways of seeing, doing and being in the world. However, the anthropologist Stephen Collier has shown that a wide array of projects can marshal critique in ways that do not necessarily advance a radical political agenda. Collier draws on his own studies on the neoliberal political theorists James Buchanan and Vincent Ostrom (Collier 2011, 2017) and Michel Foucault's (2008) examination of post-war neoliberal political and economic thought in Germany and the US in order to demonstrate how critique was an essential element in neoliberal economists' arsenal. Sweeping political and economic reforms carried out during the 1980s in the US under President Ronald Reagan and in the UK under Prime Minister Margaret Thatcher drew on decades of neoliberal critiques of Keynesian economics, which called into question not only Keynesian planners' policy prescriptions, but also the worldview of a fundamentally stable national economy that planners could guide to sustained growth and prosperity. Critique, as a practice of strategically engaging with prevailing assumptions and institutional arrangements that structure what counts as "common sense," is thus not the sole province of radical social theorists.

Collier's arguments on critique are particularly relevant for understanding the emergence of resilience. As the previous chapter detailed, ecological understandings of resilience were organized around a kind of subjugated knowledge on environmental change, a topological understanding of social and ecological dynamics that juxtaposed sharply with dominant theoretical and management approaches based on positivism and command and control, respectively. This chapter will examine how ecologists' strategic practices of critique moved resilience from the margins of ecology (and to a lesser degree, psychology and engineering) to its current status as an important organizing principle for social and environmental governance. Key here are their efforts to re-conceptualize scientific and managerial practice within a context of social and ecological complexity, which folded together ecological understandings of resilience and functionalist and behavioral understandings of governance developed in the field of new institutional economics (see Chandler 2013). As I will show, this maneuver enabled ecologists to advance a novel solution to problems of sustainable development and environmental degradation: the design of institutions that could mirror complex systems' flexibility and adaptability.

This chapter thus traces the shifting practices and strategies of critique resilience thinkers mobilized against dominant ways of thinking about and managing environmental resources. Collier's engagements with Foucault provide a roadmap for unpacking ecologists' critiques. He emphasizes the importance of the *field of adversity* that critique positions itself against. This is more than the object of critique, such as centralized planning or command-and-control environmental management strategies. Instead, a field of adversity gestures to a wider field of *strategic engagement* with the object of critique: what vision of proper forms of ordering and organizing social and environmental relations do critics promote, and *what stands in the way of realizing this vision?* In other words, a field of adversity is that element that allows critique to identify limits in established ways of doing and thinking that inhibit other ways of doing things, and show how these limits can be transgressed (Collier 2009; Foucault 2008). In what follows, I examine how resilience proponents constructed and played off their field of adversity to recalibrate research and management around topological understandings of social and ecological change. Their critical interventions advance a distinct way of knowing and engaging social and environmental phenomena organized around *both* the topological sensibilities of ecological resilience *and* what the next chapter will identify as a will to design.

BEYOND CENTRALIZATION: (MIS)ALIGNING SOCIAL AND ECOLOGICAL SYSTEMS

As Chapter 2 explained, Jeremy Walker and Melinda Cooper's (2011) genealogy of resilience has become a foundational essay for critical work on resilience. Their argument that resilience shares an "intuitive ideological fit" with neoliberalism has stimulated a growing chorus of critical research that unpacks the close affinities between resilience thinking and neoliberal governance strategies. But despite this influence – and while also noting the important role that critical work following in their path has played in denaturalizing resilience proponents' claims – Walker and Cooper's argument is perhaps best thought as a first-cut genealogical analysis that emphasizes the formal similarities between C.S. Holling's ecological theories of resilience and Friedrich Hayek's neoliberal economic thought. In brief, Walker and Cooper detail how both engage in a common critique against centralized planning and environmental management. In their own ways, Holling and Hayek both strive to show how an ontology of complexity reveals government's epistemological limitations. For his part, Holling diagnoses a key problem of the relationship between truth and the authority of government. For Holling, complex social and ecological processes exceed planning's information-gathering and -processing capacities, and thus will always confound centralized planning efforts. State efforts to design plans to maintain stable, optimal levels of ecosystem performance will inevitably lead to failure, because command-and-control management styles are constitutively incapable of understanding and controlling dynamic, interconnected complex processes.

This much is already familiar to resilience scholars. But Walker and Cooper's contribution comes from showing how the same critique of centralization motivates Hayek's work. Writing roughly thirty years before Holling, Hayek took as his object of critique the rapid expansion of state bureaucracy in many Western countries during the first decades of the twentieth century. At the time, prevailing wisdom in fields such as public administration and political science maintained that efficient government could be achieved through the centralization and professionalization of administrative services. These efforts to modernize political rule were themselves intended to address rampant corruption and "machine" politics that had held sway, particularly in the United States, for the final decades of the 1800s. They were also intended to correct for the free market's violent swings, exemplified in the Great Depression of the 1930s. But for Hayek, the rapid growth of centralized planning institutions that sought to control and direct social and economic processes in the name of efficiency instead produced inefficiencies, limited individual freedom, and set Western societies on a slippery slope towards tyranny. Key for Hayek's work was the argument that the fundamental economic activity – market exchange between two individuals – was in practice too complex for government to command (see especially Hayek 1945). Each exchange depends on a variety of unique factors that affect individual choices. This is exceedingly fine-grained information, so localized and contextually specific that an unwieldy bureaucracy could never hope to manage the informational demands required to rationally manage the allocation of resources. Instead, in Hayek's reading, only the market provides a decision-making mechanism that enables those with ready access to information – individual buyers and sellers – to make economic decisions.

Thus, for Walker and Cooper, Holling and Hayek share a form of critique that draws out the epistemic limits of government: centralized planning will lead to ecosystem and economic collapse, because decisions made through centralized planning mechanisms, as a rule, can never be based on a full and complete understanding of the complex and dynamic systems they are trying to regulate. Both diagnose a mismatch between the truth of complex social and ecological processes and prevailing governmental rationalities based on principles of centralization, efficiency and top-down, command-and-control management. However, as David Chandler (2014a) has shown in his distinction between simple and general complexity (see Chapter 2), the formal similarities end there. Hayek's critique of centralized economic planning led him to advocate for governance reforms that would allow the market to determine social and ecological outcomes. This reflects Hayek's underlying ontology of simple complexity, which maintains an external determinant of value – in this case, the so-called universal market rationality. Simple complexity posits an absolute distinction between system and world, which preserves an external point from which economists and ecologists can know and control a system with perfect certainty. For Hayek and other neoliberal economists, the absolute Truth of market rationality is unshakable and given in advance; the only challenge is how to reconfigure complex social, economic, political, cultural, and environmental interactions and practices around these market principles (Foucault 2008). In contrast, Holling's critique of command-and-control-style environmental management led him and his colleagues to advocate for adaptive management techniques that could monitor dynamic ecosystems and adjust management interventions as needed. This is not a problem of simple complexity. There is no external determinant, such as a universally valid market rationality, that can serve as the basis for judging social and ecological governance. Instead, Holling poses a problem of general complexity, which assumes that there is no position outside a complex system from which the system can be known and judged with absolute certainty. The determinants of good governance are thus immanent to the system, and the challenge becomes designing ways to constantly gather and analyze information on system performance and adapt interventions to emergent realities.

Walker and Cooper thus outline the essential elements of critical practice in both ecological resilience theory and neoliberal economic practice: they gesture towards a field of adversity – the limits centralized planning imposes on knowing and managing complex processes, whether these involve economic exchange or complex ecosystem dynamics – and advocate a common style of intervention that emphasizes decentralization, even if divergent understandings of complexity lead to different prescriptions for managing complex processes. However, the distinction Chandler draws suggests that we should be attuned to the points of *disjuncture* along with convergence between resilience theory and neoliberal governance (see also Chandler 2014b). These points of disjuncture become more apparent as we expand the genealogy of resilience beyond Holling's early work. From the 1970s to the early 2000s and beyond, a number of colleagues across ecology and related fields (especially environmental economics and human ecology) echoed Holling's call to reconfigure environmental management around the "very different view of the world" (Holling 1973: 15) that his earlier work on ecological resilience articulated. Their critiques of centralized command-and-control management strategies effectively dismantled the latter's apparent naturalness, and paved the way towards alternative management strategies based on the principles of adaptive management, institutional design, and adaptive governance. The next section explores the field of adversity that resilience proponents constructed.

TRANSGRESSING THE "GRID" OF COMMAND AND CONTROL

As the previous chapter indicated, sustainability had become a major concern for environmental managers in the 1970s. Following the publication of *The Limits to Growth*, the 1972 Stockholm conference, and the general rise of environmentalist sentiment, resource managers became gripped with Malthusian fears that the pace of economic development was outstripping the earth's capacity to supply resources that could fuel this growth. But where mainstream environmentalists and managers turned to top-down regulation to conserve resources and optimize their efficient use, heterodox ecologists instead asserted that "conventional prescriptions of resource management are in many cases not resulting in sustainability," and that "resource mismanagement and depletion" lay at the root of contemporary environmental and social problems (Berkes and Folke 1998: 1). They interpreted "limits to growth" not as an encroaching environmental horizon to be overcome through better technological development, but rather as an expression of humanity's "lagging ability" to solve resource and environmental problems (Berkes *et al.* 2003: 1). Holling succinctly summarized this view in the opening pages of his influential 1978 text on adaptive management:

> It is commonplace now to perceive limits – limits to growth, to resources, to climatic and environmental stability ... Industrial man's [sic] history, by and large, has been one of successful resolution to these problems, at least in the short term. In recent years, however, they seem to have taken the shape of crises, perhaps because the problems are ours and not our fathers' [sic]; more likely because our perceptions and methods, having once helped, now hinder.
>
> (Holling 1978: 5)

The shift here is important: the challenge to sustainable development does not simply lie in nature; instead, it lies in how humans understand and engage with nature. This inversion of the equation provided the basis for ecologists' field of adversity. The first step of introducing new ways of knowing and managing social and environmental phenomena involved detailing how prevailing "perceptions and methods," to borrow Holling's terminology, hindered rather than helped efforts to develop more sustainable resource management practices. The field of adversity ecologists constructed through these efforts consisted of two elements: the pathology of command and control, and the mismatch between institutions and the complex environments they were supposed to regulate. Each of these is considered in turn.

Field of adversity I: the pathology of command and control

As Walker and Cooper's genealogy notes, Holling's ecological understanding of resilience emerged out of what ecologists came to diagnose as the "pathology" of command and control. Command and control, as the previous chapter defined it, refers to a specific management style organized around principles of top-down, centralized regulation. It assumes systems are inherently stable and possess a single equilibrium point that demarcates the optimal use of resources. When taken up in environmental management, command and control leads to practices such as centralized state regulation of pollutants or resource extraction – practices that make up conventional resource management strategies in centralized bureaucracies around the world (Berkes and Folke 1998).

There are numerous examples of command-and-control strategies in environmental management (Holling and Meffe 1996). Pollution controls often involve some form of centralized state agency that determines legally enforceable emissions and levels fines for non-compliance. Fire management strategies rely on preventative techniques that attempt to eliminate the occurrence of fires. And strategies to promote sustainable resource use – for example, fisheries – might rely on a centralized agency that determines optimal harvesting levels and enforces compliance through licensing and surveillance. Resilience proponents acknowledge that these strategies may prove successful in the short term. Pollution levels may decrease, fire occurrence may drop to zero, and resource stocks may increase. But ecologists warn that these successes are often temporary, and may lead to larger, unexpected crises in the future. Fire management is a case in point: preventing fires in a forest allows brush and debris to build up, so that when a fire unexpectedly starts, it will be much larger and more powerful than would normally occur. This paradoxical outcome is a result of what ecologists named the "pathology of command and control:" the propensity of command-and-control management practices to transform ecosystems *and* social systems in ways that reduce adaptive capacity and make systems more vulnerable to catastrophic disruption (Holling and Meffe 1996; Folke *et al.* 2007). For ecologists, the problem is twofold: command and control transforms environments in ways that make them more prone to disruption; and it inhibits social capacities to respond to disruptions when they do occur (Westley 1995). For Carl Folke and colleagues, "the irony is that the mental model of optimal management of systems assumed to be stable and predictable has in many respects reduced options and removed the capacity of life-support systems to buffer change" (Folke *et al.* 2003: 353). Ecologists' critiques identified three specific ways command and control limited this adaptive capacity: reducing variability in social and ecological systems; separating managers from the systems they are supposed to manage; and increasing social dependence on command-and-control management practices. I want to briefly consider each limit.

First, arguably the most significant limit command and control introduced was its reduction of variability. The logic here is that reducing variability will make ecosystems "more predictable, and thus more reliable, for human needs" (Holling and Meffe 1996: 329). This war on variability is a hallmark of modernist environmental management practices (Scott 1998), and has been applied in a number of fields. Holling, writing with Gary Meffe, provides a concise list of examples:

> we control agricultural pests through herbicides and pesticides; we convert natural, multi-species, variable-aged forests into monoculture, single-aged plantations; we hunt and kill predators to produce a larger, more reliable supply of game species; we suppress fires and pest outbreaks in forests to ensure a steady lumber supply; we clear forests for pasture development and steady cattle production, and so forth.
>
> (Holling and Meffe 1996: 329)

However, for resilience scholars, this strategy is not based on careful observation of ecosystem performance. Instead, it rests on nothing more than what Fikret Berkes and colleagues identify as "the ideology of a strongly positivist resource management science, with its emphasis on command and control... [and which] aims to reduce natural variation in an effort to make an ecosystem more productive, predictable, economically efficient, and controllable" (Berkes *et al.* 2003: 8). In essence, command and control imposes an artificial order on nature that conforms with scientists' and managers' ideas (or "ideology") on how nature *should* appear and behave – an artificial order that distorts and in many cases destroys nature's inherent capacity to adapt to disturbances. In an early text, Holling (1978: 35) argues that "variability of ecological systems ... provides a kind of self-monitoring system that maintains resilience." Reducing a system's variability in the name of optimizing system performance might work in the short term, but over the long term, it "may lead to a loss of resilience in a system, leaving it more susceptible to resource and environmental crises" (Berkes *et al.* 2003: 8). Holling and Meffe (1996: 330) are even more blunt in their diagnosis of these resource and environmental crises:

> crises and surprises ... are the inevitable consequences of a command-and-control approach to renewable resource management, where it is (implicitly or explicitly) believed that humans can select one component of a self-sustaining natural system and change it to a fundamentally different configuration in which the adjusted system remains in that new configuration indefinitely without other, related changes in the larger system.

Given the disastrous effects that result from implementing command-and-control strategies, Holling (1978: 35) succinctly warns that, "policies that reduce variability in space or time, even in effort to improve environmental 'quality,' should always be questioned."

The first limit of command and control thus emphasizes the problematic effects this management strategy has on the ecosystems it is deployed to protect. The second limit focuses on the way command and control distorts the management process itself. Here, the issue is that command and control increasingly separates managers from the systems they are supposed to manage. Holling and Meffe (1996) suggest that the early success of command-and-control strategies leads to increasing bureaucratic pressures to improve on this performance, which intensifies management efforts to increase efficiency – not for improved ecosystem performance, but for bureaucratic survival. More generally, the pursuit of efficiency also leads to problematic management arrangements that attempt to compartmentalize interconnected ecosystem processes and optimize each process independently of others. Complex ecological systems are often too large for a single agency to manage on its own. Holling (1978: 48) laments the tendency for agencies in this situation to rely on consulting contracts that intensify this compartmentalization:

> Thus, a wildlife biologist might be consulted about the effects of a dam on big game animals, an economist about effects on recreation, a hydrologist about water flows, and a fisheries biologist about effects on fish... However, this approach often omits consideration of cross-disciplinary interactions, such as the effect of changing recreational demand on big-game and fish populations.

The problem here is strikingly similar to the one ecologists identified with the reduction of ecosystem variability: just as the reduction of variability imposes an artificial order on ecosystems, so too do command-and-control management strategies impose an artificial order on the relation between managers and the complex ecosystems they are supposed to manage. Rather than understanding the "cross-disciplinary interactions" that drive ecosystem change, command and control gives managers a false faith that they can know and control ecosystems simply by knowing and controlling constituent parts. However, as lessons from the collapse of the Newfoundland Atlantic Cod industry painfully illustrate (see Box 4.1), "monitoring the wrong variable can seem to indicate no change even when drastic change is imminent" (ibid.: 35). By blinding environmental managers to these system dynamics, command-and-control strategies thus inhibit environmental managers' ability to understand how the complex systems they are managing actually respond to interventions.

Box 4.1 NORTH ATLANTIC COD

The 1991–1992 collapse of the Newfoundland Atlantic Cod industry has become a paradigmatic example of command and control's myopia. Following a near collapse in the mid-1970s, in 1977 the Canadian government extended its jurisdiction over waters to 200 miles, which effectively eliminated long-distance fishing from international operations, and established a new quota system designed to optimize the sustainability of resource extraction. The centerpiece of this quota system was a revised total allowable catch calculation, set at roughly 18% of exploitable cod biomass. The utility of this calculation lay in government managers' and scientists' assumptions of a linear relation between catch rates and total stock abundance. In particular, the government's calculations showed increasing population rates, which gave it confidence that the sector was rebounding following the 1970s scare. It set quota levels accordingly (McGuire 1997).

However, on the ground, there was a growing divergence between offshore catch rates and inland catches, the latter of which were showing signs of depopulation. But inland catches came from a variety of sources, such as traps, gillnets, and handlines, that are not amenable to standardization. In contrast, government agencies relied on data from offshore trawlers, which is much more standardized and regulated, in order to calculate stock abundance and thus set catch rates.

Following the 1991–1992 collapse of the cod population, and the subsequent collapse of the regional fishing industry, observers recognized that overexploitation rather than natural causes had led to population decline (Hutchings and Myers 1994). However, they also recognized that overexploitation occurred in part because the standardized calculations government regulators used to set quotas could not capture more fine-toothed localized data that provided a more accurate picture of population dynamics. The data was there to be seen, but centralized environmental management practices, and the positivist scientific techniques that informed these practices, *prevented* managers from sensing and understanding how the population was actually behaving.

Resilience scholars took from this case the lesson that centralized, command-and-control management practices could often distort information on system performance, and thus lead to management policies that actively undermined the sustainability of the resources they were charged with managing. Effectively, the Canadian government lacked the technical and institutional capacity to incorporate data from inland operations into its management plans, and the result was a management policy based on data that state agencies could work with, rather than data that accurately represented systemic performance (Dietz *et al.* 2003).

If the first two limits ecologists identified emphasize how command-and-control management practices distort, first, ecosystems and second, managers' ability to know and manage nature, respectively, the third limit identifies how command and control distorts social practices. For ecologists, the early success of command and control leads to the widespread adoption of social practices based on the continuation of these management practices. For example, suburban development hinges on the prevention of forest fires, training rivers to eliminate flood events, industrialized and rationalized agricultural production, and so forth (Holling and Meffe 1996). The problem here is familiar: this is an artificial mode of living that dissociates individuals and communities from their natural environments, even as it increases dependence on other areas of the Earth that provide essential food, water and energy resources. Society thus becomes more tightly bound to the continued success and efficiencies of commanded-and-controlled ecosystems across the planet, even as these management strategies reduce ecosystems' variability, increase their susceptibility to disturbance, and inhibit environmental managers' ability to recognize impending crises.

Taken together, these three limits of command and control comprise one field of adversity. Each limit specifies how command and control distorts ecologists' ideal visions of scientific and environmental management practice. The first limit distorts how we can understand and value nature: command and control views nature as ontologically separate from society, and thus subject to (rationalized) social control designed to maximize efficiency through standardization and the reduction of variability – a sharp contrast to the importance resilience proponents place on redundancy and diversity. The second limit distorts management practice: command and control detaches managers from the systems they are monitoring, and thus prevents them from understanding complex interconnections. The third limit distorts social and cultural practice: command and control facilitates ways of living that are unsustainable and promotes further environmental rationalization – which further erodes resilience. Command and control thus inhibits the possibility of creating more sustainable societies because it inhibits our ability to conceptualize, manage and live with social and ecological complexity.

Field of adversity II: institutional fit

But this is not the only field of adversity resilience proponents set for themselves. In the mid-1990s, several ecologists began collaborating with scholars working in the field of new institutional economics, a branch of public choice theory (see especially Arrow *et al.* 1995; Gunderson *et al.* 1995a; Hanna *et al.* 1996; Berkes and Folke 1998). These engagements helped ecologists specify a second distinct field of adversity: the problem of institutional fit (Folke *et al.* 2007). Referencing the work of Nobel Prize-winning economist Douglass North, resilience scholars define institutions as "humanly devised constraints that shape human interaction and the way societies evolve through time" (ibid.: unpag.). Institutions encompass "formal constraints (rules, laws, constitutions), informal constraints (norms of behavior, self-imposed codes of conduct), and their enforcement characteristics" (Brown 2003: 479). Key for our purposes here, this understanding of institutions provided resilience scholars with a way to visualize the connections between complex social and ecological systems: while there was general recognition that ecological systems supplied social systems with vital resources and ecosystem services they required to survive, grow and thrive (Adger 2000), institutions mediated these relationships. They are "the mediating factor that governs the relationship between a social group and the life-support ecosystems on which it depends" (Berkes and Folke 1998: 9).

The importance of this insight for ecological work on resilience cannot be overstated. A focus on institutions enabled resilience proponents to examine change in different socio-ecological systems as outcomes of the particular institutional arrangements that shaped how social and ecological systems interacted with one another. Here, Elinor Ostrom's (1990) Nobel Prize-winning research into common property regimes was particularly impactful. In brief, Ostrom directly challenged Garrett Hardin's (1968) "tragedy of the commons" thesis, which held that in the absence of private property rights, users would over-exploit common-pool resources, or resources where exclusion of users is difficult and the benefits of consumption cannot be shared with others. In contrast, her work demonstrated "a rich diversity of common property institutions and property rights arrangements" (Berkes and Folke

1998: 8) that enabled communities to sustainably manage common-pool resources. Ostrom's arguments allowed ecologists and other resilience advocates to recognize how seemingly disconnected social and cultural practices – such as hunter's guilds, local fishery management arrangements, or even cultural taboos – provided essential ecological functions that maintained diversity and ensured continued functioning of key ecosystem processes, even in times of environmental or social stress. Thus, Ostrom's work showed that there were many possible ways of organizing the relation between social and ecological systems, and that these systems' resilience hinged on the design of institutions that governed this relation.

The multiplicity and plasticity of institutions formed the second field of adversity for ecologists: the problem of *institutional fit*. Fit refers to the "appropriateness" of institutions for achieving goals of sustainable development (Brown 2003: 479). This is a question of how institutions resonate (or dissonate) with key determinants of a system's resilience, especially functional diversity and key structuring processes that often stretch across multiple spatial and temporal scales. A lack of fit can result from any number of reasons, including "complexity and dynamics of the ecosystem, uncertainty over future changes, and factors such as irreversibility, resilience and disturbance" (ibid.: 479–480). These are all conditions endemic to systems that command and control is constitutively incapable of recognizing and engaging. Command and control can be effective in a closed system with a limited number of variables whose relationships can be known and modeled with precision. But it is far less effective in a complex, open system characterized by indeterminate relationships and emergent, self-organizing and non-linear change. Ostrom, quoting ecologist Simon Levin, suggests that, "the management of such systems presents fundamental challenges, made especially difficult by the fact that the putative controllers (humans) are essential parts of the system, and hence, essential parts of the problem" (Levin 1999, in Ostrom 2008: 28). Levin's argument raises a fundamental challenge to the "ideology of positivist science" (Berkes and Folke 1998: 8) that underpins command and control. Illustrating Chandler's (2014a) sense of general complexity, Levin asserts that there is no position outside a system from which the system can be objectively known and controlled with absolute certainty. But importantly, *the ideology of positivism enables scientists and managers to behave and act as if they inhabited such a position*. In the process, environmental managers pursue strategies and initiatives that attempt to eliminate uncertainty, increase stability, and enhance efficiency, but have the effect of reducing variability, reducing adaptive capacity and eroding resilience. The problem of institutional fit is this divergence between, on one hand, the reality of general complexity that denies the possibility of an outside, and on the other, problematic assumptions about nature, management practice and environmental change – those "formal" and "informal constraints" that shape how people (scientists and managers in particular) can visualize, understand and act within the world (see Box 4.2).

Box 4.2 INSTITUTIONAL MISFIT IN THE HIMALAYAS

In 2003, geographer Katrina Brown published an inventive critique of prevailing approaches to institutional analysis. For Brown, scholars drawing on Elinor Ostrom's design principles (see Box 4.3) tend to focus on the sociopolitical factors of institutional design, and downplay or ignore the ecological characteristics of resource systems *and* the dynamics and heterogeneity of the social itself. Instead, Brown argues that problems of instititutional fit resulted from three characteristics of conventional management approaches: first, they often have narrow or conflicting objectives; second, they tend to suppress variability and disturbance; and third, they have short-term rather than long-term goals. Problems of sustainability amplify these concerns, because sustainability brings together actors with competing interests in, for example, conservation and development. And these actors will be differentially arranged across scales, from communities to international development and conservation agencies.

Brown explores these issues through a case study of Royal Bardia National Park in Nepal, which showcases conflicts between international conservation efforts and local communities' subsistence needs. Because the region is home to many charismatic endangered species, wildlife preservation efforts had succeeded in expanding conservation areas, which had the unintended side effect of tightly restricting local people's access to vitally important forest and grassland resources. This resulted in displacement of local communities from their traditional lands. In 1994, a multi-actor partnership of national government agencies and international development and conservation groups began efforts to integrate conservation and development. However, the efforts were stymied by institutional misfit: conservationists had a narrow objective of protecting a small number of rare species (especially rhinos and tigers), which leads to a narrow set of management practices that do not reflect the diverse and dynamic interactions local communities have with their surrounding environment. Second, management approaches relied on blunt strategies such as restricting local people's access to conservation areas, which does not account for the fact that human–nature interactions have played a central role in developing and sustaining the valued environmental conditions. Third, management focused on meeting short-term ecological objectives rather than long-term "harmonization" between local peoples, protected areas, and endangered species.

In Brown's reading, institutional fit is not simply a matter of integrating diverse perspectives on complex phenomena. Instead, it is a problem that emerges out of the way contextually specific social and ecological dynamics intersect with national and global development initiatives. Thus, any attempt to address institutional misfit requires careful attention to the uneven political, social and environmental relations that gave rise to the problem in the first place.

Breaking through the "grid"

Writing in an influential paper on adaptive governance, Carl Folke and colleagues characterize ecologists' field of adversity in telling language: "the underlying worldview of resource management may *impose a grid* on social memory for managing ecosystem dynamics" (Folke *et al.* 2005: 455, emphasis added). This metaphor of the grid conveys a sense of the way institutions associated with command and control *artificially* delimit social memory, or the possible ways people might make sense out of and respond to environmental change (more on this latter term below and in Chapter 6). This grid thus inhibits the emergence of different ways of knowing and engaging social and ecological change. Fikret Berkes and colleagues echo the findings from a survey of American Association for the Advancement of Science members, which asserted that although urgent challenges facing science and society

> seemed to have radically outgrown [their] previously accepted conceptual framing'... there were new theories and explanations appearing on the horizon, many calling for more creative forms of collaboration between scientists and society, involving a broader range of disciplines and skills needed for the process.
>
> (Jasanoff *et al.* 1997, in Berkes *et al.* 2003: 1)

Resilience proponents' strategies of critique attempted to dismantle the grid of positivist science and command and control, and thus create a space for these new theories and explanations – especially the "very different view of the world" Holling (1973: 15) specified through his early studies on ecological resilience. Their formulations of the pathology of command and control and the problem of institutional fit showed command-and-control's partiality, and reduced command and control to an artificial "grid" that had been imposed on human experience of nature, not a self-evident and unproblematic environmental management approach. And because the grid had been artificially imposed, it could also be removed. In denaturalizing the claims and strategies of command and control, ecologists' critiques attempted to create a space for intervention – a space where new scientific and management practices that worked with rather than against nature could supplant established practices that no longer held purchase. The next section focuses on two interventions resilience proponents proposed in order to bring about new ways of visualizing and addressing these challenges: adaptive management and adaptive governance.

TRANSFORMING NATURE–SOCIETY RELATIONS: RESILIENCE INTERVENTIONS INTO COMPLEX SYSTEMS

In an early essay on institutional fit, Frances Westley (1995) argues that while informal institutions such as ideologies, mythologies and paradigms are essential for social action, they can also inhibit actors' ability to make sense of new or unexpected phenomena. Her argument helps us recognize how resilience scholars' engagements with new institutional economics enabled them to specify their critiques of command and control in operational terms. Specifically, it enabled them to visualize nature–society relations as a kaleidoscopic landscape of institutions that can help or hinder sustainability. A nebulous catch-all of "pathology" becomes a complex but tractable problem of institutional change. Holling and Geoff Sanderson (1996: 79, emphasis added) were clear about the critical stakes:

> human systems of property rights, built around deterministic (or stipulated) ecosystem models, are not flexible in their application or crafted in light of the temporal or spatial demands on natural systems... Until modern human institutions are built on ecological dynamism, and *designed* to flex with natural variability, their principal impact will be to neuter nature, not sustain it.

Holling and Sanderson's passage is significant not only for their diagnosis of problematic institutions, but also for the critical challenge they implicitly pose: the need to transgress the limits of command and control by *designing* different institutions. Their use of the term *design* is vitally important here, even if it remains largely unpacked in much resilience literature. As the introductory chapter suggested, the problem Holling and Sanderson pose is fundamentally a design problem: how to align an interior (human institutions) with an exterior (complex social and ecological systems) in the absence of transcendental determinants? And as the next chapter will demonstrate, this is a problem unique to general complexity and second-order cybernetics, where there is no space outside the system, no vantage point from which an external observer can render objective judgment. When humans are part of the systems they attempt to control, to borrow Levin's terminology from above, institutions are as much a part of the problem as they are the solution. Rather than blindly applying a one-size-fits-all solution (or panacea, to use one of Elinor Ostrom's (2007) terms), the challenge is to design new institutions capable of adjusting interventions to the dynamic and unpredictable world they are inextricably bound within. Resilience scholars advanced this strategy through developing practices of adaptive management and adaptive governance.

Adaptive management

A design sensibility has been an essential part of resilience scholars' critical interventions since their formative writings in the 1970s. In a landmark text, C.S. Holling (1978: 9) introduced adaptive management as a design solution that could counteract the pathology of command and control. He cautioned readers that,

> if not accompanied by an equal effort to design for uncertainty and to obtain benefits from the unexpected, the best of predictive methods will only lead to larger problems arising more quickly and more often. This view is the heart of adaptive management – an interactive process using techniques that not only reduce uncertainty but benefit from it. The goal is to develop more resilient policies.

In this passage, Holling lays out a key dimension of adaptive management that bears closer scrutiny: his assertion that "designing for uncertainty" will "develop more resilient policies" attempts to overcome the problem of institutional fit. Even though he does not use the latter term, he nonetheless is describing here how prevailing scientific institutions – the "predictive methods" of positivist science that underpin command and control – undermine resilience because they do not accommodate uncertainty. He revisits this theme more explicitly a few pages later, suggesting that, "those institutions that have developed policies that induced a rhythm of change, with periods of innovation followed by consolidation and back again, maintain a flexible and adaptive response. Expected problems and opportunities are detected and can be turned to benefits" (Holling 1978: 36). Of course, this "rhythm of change" is the movement between conservation and reorganization that characterizes what he would later define as the adaptive cycle (Holling 1986, see Chapter 3). And this is the critical intervention: designing institutions that allow management and scientific practice to mimic complex systems' inherent adaptive capacities re-introduces, in artificial form, those capacities that the pathology of command and control has steadily eroded.

This intervention takes place through at least four maneuvers: a new relation between science and management, a new relation with uncertainty, a new ethic of inductive and reflexive experimentation, and new techniques of collaborative policy design and simulation. As a way of exploring what adaptive management entails, I want to consider each of these in turn.

A new relation between science and management

For ecologists and other resilience proponents, a major limitation of command and control is its assumption that perfect knowledge of a system is both possible and desirable. This sense of omnipotence might work well for tightly bound closed systems. However, open systems present a different challenge: a complex terrain of non-linear emergence and dynamic interrelationships that confound prediction – and thus the possibility of total knowledge and control (Walker and Cooper 2011; Chandler 2014a). This requires a new understanding of the relation between science and management, one that is "based on presumptions of ignorance rather than knowledge" (Holling and Goldberg 1971). Because "what we know of social, economic and environmental behavior is much less than what we do not know" (Holling 1978: 2), "fundamental change is not simply to better mobilize known information. Rather, it is to cope with the uncertain and the unexpected. How, in short, to plan in the face of the unknown" (ibid.: 7). This is the new relation between science and management put forward by resilience proponents: a relation mediated not through presumptions of total knowledge, but rather through presumptions of limited knowledge. Science must develop ways to deal with uncertainty that its calculatory mechanisms cannot control, which introduces a degree of indeterminacy into planning and environmental management. Adaptive management replaces the hubris of omniscience with the humility of ignorance.

A new relation with uncertainty

Early proponents of adaptive management turned to the systems sciences to find techniques that would enable them to design flexible and adaptive management processes (Pelling 2010). This brought ecologists and environmental scientists into contact with research in behavioral economics and psychology influenced by Herbert Simon's understanding of bounded rationality and decision-making in complex environments. The next chapter will detail these engagements; what matters here is that this engagement provided resilience proponents with new ways to address uncertainty and thus plan in the face of the unknown. For Holling (1978: 7), "decision theory provides a few theoretical hints and some practical experience in ways to explore decisions in the face of uncertainty and conflicting objectives." This context of "uncertainty and conflicting objectives" approximates what Carl Walters (1986) refers to as a *suboptimal* situation. These are situations in which "the computational problem of finding optimal adaptive policies becomes impossible when it is impossible to specify beforehand the form of functional relationships and error probability distributions" (Walters and Hilborn 1978: 178). Simply put, a suboptimal situation is one in which scientists are unable to calculate and control for uncertainty, and thus unable to determine with certainty the optimal policy choice. However, Holling suggests that "the systems sciences have evolved methods of optimization that, if used with care, can point toward general policies that better achieve objectives by working with, rather than against, the rhythm of ecological and economic forces" (ibid.: 7). On the surface, this might seem like a contradictory claim. After all, ecologists staked out a terrain for resilience through critiquing command-and-control's efforts to optimize ecosystem performance. The difference here is subtle yet important. While command and control optimizes *outputs*, or stable states (for example, optimizing the number of fish or trees harvested), Holling and other early adaptive management scholars focus on optimizing the *process of making management decisions* within – not outside of – an emergent and unpredictable environment.

This emphasis on decision-making processes is key to understanding how adaptive management transforms scientists' relationship with uncertainty. On the first page of his field-defining text, Holling (1978: 1) suggests that, "the process of adaptive environmental management and policy design ... integrates environmental with economic and social understanding at the very beginning of the design process, in a sequence of steps during the design phase and after implementation." These steps involve:

1. Define and bound the problem;
2. Identify current knowledge of system behavior;
3. Identify uncertainties and alternative hypotheses based on evidence and experience;
4. Implement policies to manage resources while learning;
5. Monitor effects of new policies; and
6. Evaluate and reflect on lessons learned from outcomes (summarized in Rist *et al.* 2013).

Walters and Hilborn (1978: 173) characterize this decision sequence as "a process of learning about system responses through experience." This process of learning is essential because, in suboptimal situations, "the best action for any system state cannot be fixed a priori but must instead be established through sequential reassessment of system states and dynamic relationships" (ibid.). Treating the policy design process as a sequence of reflexive interventions allows scientists and managers to play with uncertainty in productive ways. If the policy intervention is appropriately sized – i.e., its failure will not result in systemic collapse – then the implementation will provoke an uncertain future into revealing itself in a controlled and manageable form. In a properly monitored intervention, observers will be able to tease out systemic connections as they reflect on the processes and relations that drove unexpected outcomes (cf. Chandler 2014a). These controlled interventions do not reveal a system's hidden truth, but rather give researchers and managers an inductive understanding of how the system responds to certain kinds of perturbations: "during the optimization process, the response of the system to different management actions must be evaluated in detail" (Walters and Hilborn 1978: 169). They can then feed this knowledge into the next iteration of the adaptive management process, and adjust their policy intervention accordingly.

Walters and Hilborn (1978: 183) assert that "the whole process of adaptive management may be viewed as a problem of sequential experimental design, with the nasty twist that every experiment must be evaluated in terms of its contribution to the future stream of system states and payoffs." This "nasty twist" is what allows scholars and researchers to engage with uncertainty in new ways. The practice of evaluating unexpected outcomes and adjusting future interventions accordingly transforms uncertainty: no longer is it a disruptive force that prevents optimization; it is now a vital resource that drives forward the decision-making process.

A new ethic of experimental learning

When uncertainty is unavoidable, presumptions of total – or even sufficient – knowledge lead to management policies that "become more the source of the problem than the solution" (Holling and Dantzig 1976: 3). Walters and Hilborn (1978: 157, emphasis added) remind us that,

> the practice in fields such as fisheries management has often been to develop deterministic prediction models based on the best available estimates of dynamic parameters, then to hedge against uncertainty by adopting somewhat more conservative behavior than the models predict to be optimal. While the pretense is scientific management, *mistakes and failures are seldom treated as useful adaptive experiments or tests of understanding: we bury our mistakes instead of learning from them.*

The italicized passage signals a distinct ethic that guides the adaptive management process: rather than treating policy implementation as a top-down application of infallible expert knowledge, it is a chance to demonstrate knowledge's blind spots and learn from mistakes that result from these limitations. Optimizing the policy design *process* – not output – enables ongoing learning and adjustment, as scientists and managers adapt both their understandings and their policy choices to an emergent context. Indeed, for Walters and Hilborn (1978: 173), managers and scientists have little choice but to adopt this experimental outlook: "ecological management in the foreseeable future must rely to some extent on adaptive policies of learning by doing, and we should be concerned about how to optimize this adaptive process."

Optimizing this adaptive process entails transforming key components of prevailing research and management practices. In a review of adaptive management literature, Carl Folke and colleagues suggest that,

> because the self-organising properties of complex ecosystems and associated management systems seem to cause uncertainty to grow over time, understanding should be continuously updated and adjusted, and each management action viewed as an opportunity to further learn how to adapt to changing circumstances. This is the foundation for adaptive management wherein *policies become hypotheses, and management actions become experiments to test hypotheses.*
> (Folke *et al.* 2005: 447)

Adaptive management blurs the divide between management and research: in the absence of certainty, any policy intervention's effects are unknown in advance. Thus, an intervention is a kind of hypothesis that tests not only our expectations for the outcome, but also our current, limited understanding of system dynamics that give us this expectation. With limited understanding of system dynamics, scientists and managers must collaboratively develop a range of alternative hypotheses about ecosystem performance, and a corresponding evaluation of experimental policy options. Holling (1978: 15–16) reminds us that inductively exploring different policy options is particularly important given how, in the absence of certainty, "there may be many ways of attempting to achieve a given objective. For example, maximum sustained yields from a fishery can be reached by controlling fishing effort through manipulation of open fishing days or by setting catch quotas." Given this topological plasticity, the question is no longer "what is the optimal policy choice?" but rather, "how might different policy choices impact future system states and policy options?" – a question that requires ongoing, reflexive and experimental engagement with the system rather than the aloof detachment the pathology of command and control facilitates.

New techniques: collaborations and simulations

The various transformations and inversions identified thus far all upend the modernist privilege of science as the sole arbiter of objective truth. As Herbert Simon (1955) recognized with his concept of bounded rationality (detailed in the next chapter), total knowledge of complex systems is impossible. Dealing with complexity instead requires working across the artificial boundaries modernism constructs across the scientific disciplines and between science and management. The goal here is not to construct an objective and total understanding, but to synthesize different visions of what the system encompasses, how it operates, what problems need to be addressed, and how they can be addressed. Adaptive management provides one way to synthesize different kinds of knowledge through the collaborative policy design process. For proponents of adaptive management, collaboration is a specific technique for managing complexity in open, indeterminate systems. It refers to a specific kind of cross-boundary interaction: that between scientists and environmental managers. It is not a one-size-fits-all panacea for the challenges complexity poses, but instead is context dependent: its precise nature will be flexible and determined by the particular issues and projects it seeks to address. Moreover, since collaboration will involve scientists and non-scientists, the purpose of collaboration cannot be reduced to the scientific pursuit of knowledge. Holling (1978: 37) reminds us that while experiments designed to produce information can be part of a research plan, "others should be designed into actual management activities. Managers as well as scientists learn from change."

As a technique for managing complexity, collaboration intervenes at a very particular moment in the policy design process. Holling (1978: 48) describes collaborative workshops as "the core of adaptive assessment." In his view, collaboration for adaptive management is a process that "depends upon a small group of people that interacts with a wider set of experts during a series of short-term, intensive workshops" (ibid.: 49). The benefit of these workshops is that bringing together a small group of managers and researchers from different disciplinary backgrounds "circumvents the scientist's natural tendency to break problems down into components, and those components down into subcomponents, and so on," a tendency that "is often not suitable for dealing with management concerns that are at a different level from those of the scientist ... and that are likely to lie between usual areas of disciplinary interest and training" (ibid.). But for Holling, the collaborations still have a precise purpose: "most of our workshops have used the construction of a quantitative model as a focus for discussion, but as we will demonstrate later, many benefits will arise from workshops even if other predictive methods are substituted." The purpose of collaboration is thus to combine different understandings of how a particular system might respond to a disturbance. Faced with a suboptimal situation, synthesizing this multiplicity of views in the assessment process produces a complex understanding of system behavior that mimics the complex system dynamics management targets.

For Holling and other early advocates of systems-based approaches to ecology and resilience, this understanding took the form of models that simulate, but do not represent, systemic dynamics. The difference here is subtle but important. Representation adheres to the correspondence principle, which assumes that one entity, such as an abstract mathematical formula, can stand in for another entity, such as a biochemical process. In positivist natural and social sciences, representation assumes that physical and social phenomena can be objectively represented – and thus known, predicted, and controlled – through mathematical calculation (Gregory 1978; Elden 2006). Representation, and the predictive knowledge it produces, is a core assumption behind command-and-control management strategies: science determines the appropriate representation of the processes management seeks to control, which allows science to inform management of the optimal policy choice. In contrast, simulation does not produce objective, quantitative knowledge. Instead, it produces qualitative knowledge about the relations between elements that comprise a system. In language from the previous chapter, these are topological relations; thus, in the case of environmental management, simulation generates qualitative knowledge about the relations between a diversity of species and the stability of the ecosystem they inhabit. For Kai Lee (1993: 57), whose important work on adaptive management we will consider in the next chapter, "seeing the ecosystem as a whole must precede efforts to manage it." Collaborative model design is thus a vital technique of adaptive management. Holling (1978: 6) describes the characteristics and uses of simulation models in language that is worth unpacking:

> Systems ecology, in partnership with the physical sciences, has now matured enough to be capable of producing succinct representations of key elements in social and environmental systems. The resulting models mimic not simply static properties, but the dynamic ones that shift and change because of natural and man-induced influences. They can serve, alone or combined with similar economic representations, as a kind of laboratory world for the development of alternative policies and for the exploration of their impact.

Holling's use of the phrase "laboratory world" is particularly important here. The point is not to create a model that fully represents, in one-to-one correspondence, the complex system in question. While more detailed models may be more accurate, they are not useful for management purposes. Walters and Hilborn (1978: 160) warn that, while "a realistic model of most populations would keep track of all age or size classes, as well as predators, food and perhaps spatial distribution," this detail may not be useful to the policy design process because "such complex models are much too detailed for those existing optimization techniques that provide feedback policy solutions."

At issue here is how breadth rather than depth is required for answering complex management questions (Holling 1978). Hence the use of computer modeling developed in the systems sciences, which "mimics behavior over time for a variety of conditions." For Lee (1993: 62), models enable both scientists and managers to examine how "assumed behavior and data fit together" while recognizing that both our assumptions about system behavior and data are necessarily partial and limited. This does not require accurate representation, but instead demands that models "provide a consistent framework for comparing alternative courses of action" (ibid.: 61). And as Lee stresses, "'consistent' is not the same thing as 'objective.'"

Thus, in its early formulations, adaptive management depended on collaborative actions between scientists and managers that could produce "a well-validated simulation model that could be used as a laboratory world to aid in the design and evaluation of alternative policies."

A model that approximates – but does not represent – system behavior allows both scientists and managers to enact, in an artificial and controlled form, "experiments" on complex systems. Scientists and managers can test the hypothesis of a policy intervention in the model's simulated "laboratory world," and they can feed the outcomes of this simulated intervention back into the adaptive management process in order to compare the results to established understandings of system behavior and desirable policy outcomes. If a reliable model consistently generates undesirable outcomes from a given policy, then scientists and managers can be comfortable – but not certain – in pursuing alternative policy choices.

The limits of adaptive management

Taken together, this new relation between science and management, new understanding of uncertainty, new experimental ethos, and new techniques of collaboration and simulation subvert the pathology of command and control and the problem of institutional fit. Adaptive management attempts to introduce adaptability into complex socio-ecological systems through creating a new process of producing knowledge on system dynamics. In the process, it reconfigures the relation between science, knowledge, management, and the complex reality management seeks to govern. Scientists become one of many actors with an interest in environmental performance, and their traditional pursuit of objective knowledge is transfigured into the collaborative development of practical, pragmatic knowledge tailored to specific management problems in specific contexts (Holling 1978; Lee 1993). Management no longer involves the direct, top-down application of scientific knowledge, but instead involves a collaborative process of policy design and assessment that synthesizes multiple perspectives in each iteration of the design process. The result is flexible and adaptive policy that mimics the flexibility and adaptability of the system it seeks to govern.

Of course, this analysis here is overly formal and diagrammatic: in practice, the success of adaptive management is much less straightforward (Walters 1997; Lee 1999; Rist *et al.* 2013). Resilience scholars and sympathizers began encountering a number of difficulties as they attempted to apply the foundational work of Holling, Walters, Lee and others. Walters (1997) himself suggested that existing institutions and entrenched interests tend to see adaptive management as a threat to their own interests. On one level, this implicitly recognizes Paul Nadasdy's (2010) claim that the emergence of resilience is part of ongoing processes of conflict and struggle over understanding and managing social and environmental change, rather than a simple paradigm shift. However, many resilience scholars drew a different lesson from Walter's observation. They increasingly asserted that adaptive management focused "only on the ecological side as a basis for decision making for sustainability," and left out the social side of the equation (Folke *et al.* 2005: 443). The result was an overly narrow understanding of resilience that did not take into account the influence of social system dynamics on broader socio-ecological complexity. Thus, beginning in the late 1990s and extending to the present day, resilience scholars increasingly focused on what Per Olsson and colleagues (2006: unpag.) refer to as "transformations within the social domain of the [socio-ecological systems] that increase our capacity to learn from, respond to, and manage environmental feedback from dynamic ecosystems." These transformations include a laundry list of cultural and institutional factors, including "shifts in social features such as perception and meaning, network configurations, social coordination, and associated institutional arrangements and organizational structures... [as well as] redirecting governance into restoring, sustaining and developing the capacity of ecosystems to generate essential services" (ibid.). This shift in the terrain of intervention, from the ecological to the social, facilitated the emergence of a second form of critical intervention into prevailing environmental management practices: adaptive governance.

Adaptive governance

Carl Folke and colleagues (2010: unpag.) proclaim that "extending the use of resilience to social-ecological systems... is an exciting area of explorative work broadening the scope from adaptive management of ecosystem feedbacks to understanding and accounting for the social dimension that creates barriers or bridges for ecosystem stewardship of dynamic landscapes and seascapes in times of change." Their emphasis on *broadening* adaptive management is key to understanding what adaptive governance entails. It does not supplant adaptive management techniques; rather, it pushes in new directions adaptive management's efforts to transgress the limits of the pathology of command and control and the problem of institutional fit. In Folke and colleagues' (2005:

444) terminology, adaptive governance "expands the focus from adaptive management of ecosystems to address the broader social contexts that enable ecosystem-based management." Governance here refers to institutional arrangements that "structur[e] processes by which people in societies make decisions and share power," which in turn shape individual and collective actions (ibid.; see also Lebel *et al.* 2006). While this operational definition contains problematic assumptions about the nature of power, which Chapter 6 will unpack, what matters here is that a focus on governance gives resilience scholars a conceptual foothold to critically intervene in institutions that mediate relations between social and ecological systems. If, as Louis Lebel and colleagues (2006: unpag.) argue, "a society's ability to manage resilience resides in actors, social networks, and institutions," then working on governance arrangements allows resilience proponents to design interventions that will increase social capacity to manage resilience.

Just as resilience proponents' formulation of the problem of institutional fit hinged on adapting Elinor Ostrom's work on common property regimes to the pathology of command and control, so too does adaptive governance hinge on adapting Ostrom's work to the series of inversions and technical developments that comprise adaptive management interventions (Berkes and Folke 1998; Berkes *et al.* 2003). One of Ostrom's major contributions was her specification of eight "design principles" meant to characterize recurring features of common-pool resource management arrangements that had persisted for long periods of time (see Box 4.3). Importantly, these design principles were meant to counter the tendency to rely on one-size-fits-all panaceas, such as market-based solutions or state-led centralized planning. Rather than describing an ideal governance arrangement that could be systematically applied in a top-down manner across different contexts, her work showed how different techniques and strategies – or "principles," in her terminology – could be pragmatically re/configured and synthesized in unique ways to meet specific challenges of resource access or service provision. While Ostrom herself adapted these principles to questions of sustainability and resilience (Ostrom 2008), other resilience scholars took up both these principles and Ostrom's underlying pragmatic and synthetic approach to institutional design. The result was a series of transformations in the interventions they pursued under the banner of adaptive management. Each transformation is considered below.

Box 4.3 ELINOR OSTROM'S GENERAL PRINCIPLES OF INSTITUTIONAL DESIGN

Elinor Ostrom's (1990) Nobel Prize-winning work provided a sharp critique of established debates on the governance of common-pool resources, such as water supply or fish populations. Ostrom targeted what she saw to be a problematic tendency, on both the political right and political left, to rely on top-down, one-size-fits-all governance models. On one hand, followers of Garrett Hardin's (1968) "tragedy of the commons" thesis asserted that common-pool resource systems were doomed to collapse in the absence of private property rights; while on the other hand, left-leaning scholars and activists tended to argue for centralized state control in order to regulate resource use in a more rational and just manner. For Ostrom, these strategies both relied on one-size-fits-all models of resource use that failed to engage with the complexities of contextually specific resource management problems. Privatization and centralization *both* had the potential to distort information and lead to management activities that would exacerbate rather than solve resource shortages. Instead, she argued that local peoples had frequently developed successful ways of managing common-pool resources *without* relying on either the market or the State. However, the solutions they developed were contextually specific, and did not lend themselves to generalizable laws that could provide an objective guideline for institutional design in other locations.

To account for these successful management activities, Ostrom proposed a series of design principles intended to signal the "raw ingredients" of successful common-pool resource governance. The precise configuration of these principles is contextually dependent, but Ostrom's comparative case studies suggested that successful management typically involved some combination of the following:

1. *Clearly-defined boundaries* that delimit the extent of the resource, the appropriate users, and their rights
2. *Congruence between rules of use and local conditions* that insure rules of appropriation and provision are aligned with local

social and environmental conditions (roughly analogous to resilience proponents' problem of institutional fit)
3. *Collective-choice arrangements* that allow users to participate in modifying rules of use
4. *Monitoring* of resource conditions and resource user behavior
5. *Graduated sanctions* that increase the penalty of rule violation as violations occur
6. *Conflict-resolution mechanisms* that allow users and officials to resolve conflicts
7. *Minimal recognition of the right to organize*, which allows users to develop their own institutions without external challenge
8. In larger systems, *nested enterprises* allow multiple, overlapping organizations across a number of scales to carry out these functions

These principles function as a heuristic device for thinking through the different ways that institutions, as rules that order social relationships, shape particular decision situations. These include boundary rules (who can use a resource), scope rules (matching rules to the physical world), position and authority rules (prescribed roles for different actors, and ensuring outside authorities respect community rule-making capacities), aggregation rules (how to determine resource use), procedure rules (rules for monitoring others' behavior and managing conflicts) and information rules (communication procedures among resource uses) (see Kiser and Ostrom 1982: 194–195). Disaggregating institutional design into these components allows scholars to better identify specific and targeted governance and policy interventions that can improve environmental management outcomes while accounting for conflicts and issues that inevitably arise in complex social and ecological systems.

Transforming the new relation between science and management

As we saw above, adaptive management recast the relation between science and management within a context of non-linear emergence and incalculable uncertainty. Rather than science providing management with expert knowledge to direct decision-making, both science and management had to recognize and deal with the limits social and ecological complexity imposed on their knowledge. They deployed techniques such as adaptive management, science–policy collaborations and simulations in order to plan from a position of ignorance rather than certainty. However, resilience scholars' engagements with new institutional economics added another layer of complexity. According to Carl Folke and colleagues, "facing complex adaptive systems and periods of rapid change gives the scientists a new role in decision-making from being an objective and detached specialist expected to deliver knowledge to managers to becoming *one of several actors* in the learning and knowledge generation process" (Folke *et al.* 2005: 445, emphasis added). An important transformation is evident here. Managing complex systems through the process of "learning and knowledge generation" does not simply involve scientists and resource managers; it now includes knowledge and interests from a variety of different social actors, such as community groups or, in many cases, indigenous peoples. This transformation further decenters scientific knowledge, and forces scientists to engage with a variety of other forms of knowledge and interests.

Transforming the new relation with uncertainty

For Berkes and Folke (1998: 14), "traditional systems [of resource management and environmental knowledge] represent many millennia of human experience with environmental management, and provide a reservoir of active adaptations which may be of universal importance in designing for sustainability." Ostrom's (1990) design principles helped resilience scholars conceptualize these different forms of knowledge as culturally and geographically specific ways of dealing with complex uncertainties (Berkes and Folke 1998; Berkes *et al.* 2003). These distinct forms of knowledge could also contribute to more general understandings of how to live with dynamic social and ecological systems. Folke and colleagues suggest that, "many societies interpret and respond to feedback from complex adaptive ecosystems, and their practices are not only locally adaptive, but generalizable over wide regions. These practices provide insights for contemporary resource management, and complement existing approaches" (Folke *et al.* 2003: 374). Such insights provided resilience scholars with new understandings of how to manage uncertainty: designing ways to synthesize these diverse knowledges and experiences. By 2005, Folke and colleagues could reflect that "efforts are taking place to mobilize, make use of, and combine different knowledge systems and learning environments to enhance the capacity for dealing with complex adaptive systems and uncertainty" (Folke *et al.* 2005: 445). Here, ecologists also borrowed the concept of knowledge system from science and technology studies (e.g., Watson-Verran and Turnbull 1995) to signal the "institutional and social context" that fosters "learning how to sustain social-ecological systems in a world of continuous change" (Folke *et al.* 2003: 373). Reading "knowledge systems" alongside Ostrom's design principles, resilience scholars thus identified a novel way to approach uncertainty: exploring "the potential in combining local knowledge systems with scientific knowledge to cope with change in resource and ecosystem management" (Folke *et al.* 2005: 446).

Transforming the new ethic of experimental learning

Resilience scholars' newfound imperative to synthesize distinct forms of knowledge transformed adaptive management's ethic of experimental learning. It shifted the locus of experimentation and learning from the interface between science and system – that is, the knowledge of the system itself – to the institutional context that shaped the possibilities for learning within complex social systems. In other words, the challenge was no longer to devise adaptive and reflexive techniques to gain better knowledge about complex systems, but to understand the so-called "social sources of resilience" that enable adaptive management and reflexive learning in the first place (Folke *et al.* 2005). These sources of resilience all point to various institutional constraints and factors that can inhibit or facilitate learning: learning how to live with rather than resist change; combining and synthesizing different knowledge systems, rather than adjudicating between them; creating opportunities for self-organization rather than insisting on top-down command and control; and nurturing the sources of resilience rather than eliminating variability, diversity and redundancy.

These four sources encompass resilience proponents' critiques of the pathology of command and control and the problem of institutional fit, but they do so in a way that shifts the terrain of intervention to the processes and relations that shape the behavior of social systems. Discussing the concept of social learning, Folke and colleagues suggest that "social systems are structured not only by rules, positions and resources but also by meaning and by the entire network of communicating individuals and organizations at different levels of interactions, representing the social system involved in governance of ecosystems" (Folke *et al.* 2005: 448). This passage reveals an implicit assumption that social systems are essentially information-processing arrangements, a link with cybernetics research whose ethical and political implications Chapters 5 and 6 will unpack. But it also signals how resilience scholars increasingly recognized institutional fit as a question of the distinct processes that generate meaning and learning. Folke and colleagues go on to suggest that "learning that helps develop adaptive expertise … and processes of sense making are essential features of governance of complex social-ecological systems, and these skills prepare managers for uncertainty and surprise" (Folke *et al.* 2005: 447). Here, a new formula for adaptation emerges. Adaptation involves activating that diversity of past experiences and knowledges that might not register as properly scientific but nonetheless contribute to understanding and managing change. But existing informal institutions such as ideologies, mythologies and paradigms can inhibit social capacity for resilience, for they can impose meaning on different peoples, knowledges, ecological processes and outcomes in ways that prevent individuals' and organizations' ability from making sense out of new or unexpected phenomena (Westley 1995).

Here, resilience proponents increasingly alighted on social networks as key to understanding and overcoming institutional constraints on learning. Networks refer to informal governance systems across and between different scales. They provide researchers with one way to operationalize social capital, since a focus on building relationships within communities and between communities and agencies and organizations at higher scales can increase trust and facilitate communication across scales. Effective communication across scales is essential for learning, since local-scale institutions can learn, adapt and respond to environmental feedbacks and surprises faster than centralized state agencies (Westley 1995; Folke *et al.* 2007). But consistent with Ostrom's (2009) wariness of panaceas, this does not result in venerations of the local as the solution to complexity. Instead, resilience proponents emphasize the need for complex, nested, polycentric institutional arrangements capable of linking the centralized state's ability to command resources with local-scale networks' abilities to learn and rapidly adapt to emergent conditions (Tompkins *et al.* 2002; Dietz *et al.* 2003; Anderies *et al.* 2004).

Thus, the challenge for resilience scholars becomes designing networks that can facilitate cross-scale and cross-boundary linkages. This brings us back to the problem of experimentation and learning. Recall from above that adaptive management envisions policies as hypotheses and management actions as experiments to test hypotheses (Folke *et al.* 2005). When the level of analysis shifts to designing the institutional context in which decisions on policies are made, the object of experimentation likewise shifts: what is being tested are not policies through management actions, but the efficacy of different institutional arrangements to develop and support reflexive and adaptive policy design – i.e., the social capacity for resilience. And experimentation no longer occurs through controlled interventions that test the limits of scientific and management knowledge on complex systems. Instead, it involves developing comparative case studies that provide lessons learned on the factors that develop or inhibit this social capacity (Olsson *et al.* 2006; Olsson *et al.* 2014). The subtle shift in the practice and logic of "experimentation" reflects how the question of institutional design is indeterminate – by its very nature, design recognizes that there is no transcendental determinate of value that can provide an objective basis for judging a design's appropriateness. And given the large number of possible combinations and re/combinations of design principles, each management situation will be unique – it will involve a distinct array of issues, interests and actors that defies easy categorization. The challenge thus becomes identifying and reflecting on what did and did not work in a given situation (Olsson *et al.* 2006). This reflection produces inductive knowledge on the formal and informal institutional factors that enhanced or inhibited resilience in the specific case study.

Transforming techniques of intervention: co-management and transformation

If adaptive management placed new demands on the process of managing and generating knowledge about complex ecosystems, adaptive governance places new demands on the governance system that forms the institutional context in which management and knowledge generation occurs. As Folke and colleagues note, "to emerge and be effective, self-organized governance systems for ecosystem management require a civic society with a certain level of social capital, and the governance system must continuously learn and generate experience about ecosystem dynamics" (Folke *et al.* 2005: 452). Moreover, Anderies and colleagues suggest that, "to enhance the robustness of [social and ecological systems], it might be desirable to have institutions that are not persistent but may change as social and ecological variables change" (Anderies *et al.* 2004: unpag.). These challenges stake out adaptive governance's terrain of intervention. Adaptive governance relies on at least two techniques to spark changes that facilitate learning and flexibility at the level of institutional context: co-management and transitions management.

First, the concept of co-management (or adaptive co-management) extends the collaborative ethos of adaptive management to the wider social context. If, as suggested above, a variety of social actors have valid and valuable experience with and knowledge about managing uncertainty, then the problem becomes how to effectively integrate this knowledge into the planning process. Co-management provides one solution. It "explicitly recognizes the necessity of combining adaptive management with institutions across scales" (Folke *et al.* 2003: 375). This effectively attempts to create institutional arrangements that bring together local stakeholders and other interest groups with decision-makers at larger scales. In theory, these networks can design management plans that take into account diverse forms of knowledge and interests and decentralize decision-making responsibilities and resources. Proponents maintain that, to the extent that these arrangements give local groups most directly involved with resource systems more influence over decision-making processes, co-management can combat the pathology of command and control. Its multi-scalar and adaptive structure also overcomes the problem of institutional fit. Folke and colleagues suggest that, "adaptive co-management benefits from combining the ecological knowledge of local resource users, scientists and other interest groups, often with different conceptualizations of these issues and even different worldviews and belief systems, in mutual learning systems" (Folke *et al.* 2003: 380). These "mutual learning systems" are design interventions that synthesize distinct forms of knowledge and distinct understandings of problems into tractable and adaptive management plans to create more resilient social and ecological systems.

On paper, co-management looks enticing. However, as Paul Nadasdy's (2003) ethnographic study of the Ruby Range Steering Committee in the Yukon demonstrated (detailed in the introduction to Chapter 3), it can often reinforce rather than challenge existing political economic and cultural inequalities. In part this hinges on the implied assumption that these different forms of knowledge and worldviews can be synthesized, or translated across difference in ways that produce a unified and coherent "whole" of a collaboratively designed management plan (Nadasdy 1999). This gave rise to a variety of critiques from both sympathetic and more critical scholars, who argued that in practice resilience offered little more than cover for maintaining the existing status quo in the face of social and ecological uncertainty (e.g., Davidson 2010; Pelling 2010). And as we saw in Chapter 2, other scholars argued that co-management's emphasis on "therapeutic" interventions designed to build trust and solidify social networks diverted resources and attention from more pressing issues such as persistent structural inequalities that left marginalized peoples more vulnerable than others to social and ecological change (Gaillard 2010; Cannon and Müller-Mahn 2010).

These critiques led resilience scholars to pursue a distinct form of intervention: transformation. Per Olsson and colleagues define transformation in the following terms:

> transformability means defining and creating novel system configurations by introducing new components and ways of governing [social and ecological systems], thereby changing the state variables, and often the scales of key cycles, that define the system. Transformations fundamentally change the structures and processes that alternative feedback loops in [social and ecologocial systems].
>
> (Olsson *et al.* 2006: unpag.)

This definition draws inspiration from the transitions management literature in science and technology studies (Rip and Kemp 1998; Smith *et al.* 2005; Smith and Stirling 2010). The confluence between resilience thinking and transitions management enables resilience scholars to systematically unpack how change occurs in social and ecological systems (Folke *et al.* 2010; Olsson *et al.* 2014). Most importantly, it also facilitated a decisive break from their increasingly problematic conceptual reliance on adaptation. If adaptation emphasized developing a system's capacities to remain within a basin of attraction despite undergoing transformation and change (see Chapter 3), transformability means "creating and defining a new attractor that directs the development of the social-ecological system" (Folke *et al.* 2005: 457). Importantly, because ecological systems' complexity poses a barrier to calculated human intervention, the kinds of transformations they describe all target the social side of complex social-ecological systems. Folke and colleagues suggest that, "transformational change often involves shifts in perception and meaning, social network configurations, patterns of interactions among actors including leadership and political and power relations, and associated organizational and institutional arrangements" (Folke *et al.* 2010: unpag.).

This understanding of transformability creates a new logic of intervention: the possibility that institutional change can be strategically targeted in order to open up new development pathways for the system in question. According to Olsson and colleagues,

> Three main phases of transformations in social-ecological systems have been identified: (1) preparing for transformation, (2) navigating the transition, and (3) building the resilience of the new direction. The first and second phases tend to be linked by a window of opportunity. In the preparation phase, agents of change and their networks may work simultaneously at different scales of the social-ecological system. By intervening at broader institutional levels, they can open up new trajectories of development.
>
> (Olsson *et al.* 2014: unpag.)

Olsson's formulation opens up two key areas of concern. First, part of preparing for transformation requires resilience scholars to understand when interventions might be most appropriate and effective. Accordingly, some researchers have begun focusing on so-called "socio-ecological traps," which Joshua Cinner (2011: 835) defines as "situations when feedbacks between social and ecological systems lead toward an undesirable state that may be difficult or impossible to reverse." Understanding how these traps emerge through comparative case study analysis enables scholars to identify problematic feedback loops and determine points in that loop where intervention might transform social and ecological systems into new and more desirable states (see Box 4.4).

Box 4.4 RESILIENCE TRAPS IN THE FLORIDA EVERGLADES

The Florida Everglades are one of the world's largest wetlands and a unique ecosystem. For the past one hundred years, it has also been a site of intensive land use change, through the introduction of industrial-scale agriculture and expansive urban and suburban development. It is thus a space that folds together multiple and competing interests, even as conservationists have succeeded in winning governmental support for restoration efforts. However, Lance Gunderson and colleagues (2014: 155) suggest that "entrenched organizational hierarchies" have limited the success of environmental governance in the region. By this, they mean that established governance networks and management strategies undermine well-meaning efforts to restore the Everglades after decades of environmental degradation. Even though environmental management in the region has been met with periodic crises since the early 1900s (Light *et al.* 1995), the solutions to these crises often simply intensify failed management strategies: "responses to perceived ecological crises have been large-scale, extensive, and technologically based solutions: more money, more concrete, more control" (Gunderson *et al.* 2014: 155). Entrenched interests often prevent more experimental and adaptive environmental management strategies. This is particularly the case with powerful agricultural interests in the region, such as sugar cane growers. Sugar farming on the industrial scale practiced in the Everglades requires large quantities of water as well as intensive use of fertilizers. Channeling water into agricultural use thus starves other parts of the region. This loss of water couples with fertilizer run-off to make sugar farming a major culprit in Everglades degradation (Hollander 2012). However, state and federal restoration efforts continue to guarantee farmers access to the same amounts of water (Zellmer and Gunderson 2009), which reduces the possibility for unconventional restoration approaches. The Everglades thus illustrates how the resilience of established political economic interests and prevailing environmental management strategies inhibit the development of more sustainable social and ecological futures.

Second, the need for strategic intervention also raises the question of what kinds of interventions might bring about the kinds of transformations resilience proponents advocate. Again, there is no one-size-fits-all model. As Olsson and colleagues argue, "the problem is not that knowledge is coming from the outside, but rather than people are obliged to adapt to a standard 'solution' rather than the solution being adapted to specific ecological, social and cultural conditions" (Olsson *et al.* 2014: unpag.). Their formulation presents a design challenge to resilience scholars: there are a variety of innovations such as various subsidies, "new governance modes, business models, microcredits and crowd sourcing" (ibid.), which can all provide actors with different incentives to change livelihood strategies or resource use, but the precise configuration – or rather, synthesis – that will be effective and accepted in particular contexts must be worked out in place. Thus, building resilience requires what Folke and colleagues describe as "creating social and institutional space or platforms for dialogue and innovation," which can "stimulat[e] learning and resolv[e] social uncertainties" (Folke *et al.* 2003: 360). While they were explicitly discussing the possibilities for post-crisis resilience interventions, their concern with developing "platforms" has carried over into a broader concern with "the role of interactive innovation spaces and change labs within which farmers, researchers, businesses, policy makers, and others can co-produce solutions" (Olsson *et al.* 2014: unpag.). Consistent with transitions management literature on niches (Smith and Raven 2012) or literature on shadow networks for climate change policy development (Pelling 2010), these are spaces where new innovations, and potential combinations between different innovations, can be tested in relative shelter from wider social and political economic pressures that would otherwise derail their development. In effect, the creation of innovation spaces allows researchers, in collaboration with other interest groups, to experiment with different solutions to particular problems of social and ecological complexity – solutions which can stretch from highly specific technological developments to new institutional arrangements to guide decision-making processes.

CONCLUSIONS: DESIGNING RESILIENT SOCIAL AND ECOLOGICAL SYSTEMS

The convergence between ecological resilience and institutional analysis thus created a new problem space for critical intervention: the institutional context in which environmental management occurs. Co-management and transformability provide resilience proponents with two distinct ways of intervening in this wider context in order to develop adaptive governance arrangements. Each moves beyond the narrow focus of previous critical interventions. If adaptive management collaborations and simulations enabled resilience scholars to generate new kinds of knowledge about complex systems and their management that departed from positions of ignorance rather than certainty, techniques of adaptive governance produce inductive and empirically grounded understandings of the way formal and informal institutions inhibit social capacities for adaptation, and create ways to explore how these constraints might be transformed or transgressed. However, taken together, these techniques enabled resilience scholars to engage in critical interventions against prevailing environmental management techniques on two levels: the level of knowledge production and policy choice, and the level of institutional design that shapes the rules and norms that implicitly shape knowledge production and policy choice. Although not without problems (Nadasdy 1999; Voß and Bornemann 2011), these interventions have challenged the dominant paradigms of environmental management that many scholars and managers simply took for granted. And by the first decade of the new century, this new way of understanding social and ecological change has become mainstreamed into a variety of academic and professional fields (Chandler 2014a).

Mark Pelling (2010: 30) suggests that "[adaptive management's] major contribution is in taking us from abstract, modeling or conceptual work to that based firmly in the empirical reality of decision-makers who wish to mainstream adaptation into changing socio-ecological contexts." As this chapter has argued, this process of "mainstreaming adaptation" occurred through resilience proponents' sustained practice of critique, which drew out the limits of established command-and-control-based environmental management strategies and proposed a series of interventions that would transform these strategies. The resulting adaptive management and adaptive governance techniques reconfigure both social and environmental management around the topological sensibility of ecological resilience – in theory if not always in practice (Walters 1997; Rist *et al.* 2013). As Chandler (2014a) details, this sensibility underpins a number of subtle but important transformations in governance rationalities that reconfigure how governance occurs within a complex world. Foremost among these is Kai Lee's assertion, borrowed from ecologist Aldo Leopold, that managing complex systems requires a "change in the role of Homo Sapiens from conqueror of the land-community to member and citizen of it" (Lee 1993: 54). Lee recognizes that there is no longer a split between humans and the world that affords the objective certainty that comes from a "god's-eye view." Instead, as part of complex social and environmental systems, the challenge for managers and scientists alike is to invent – or more properly, design – new ways for gaining knowledge on and intervening in these complex relations.

Resilience scholars' practices of critique deftly synthesized seemingly distinct bodies of knowledge – such as complex systems theory, ecology, new institutional economics, science and technology studies and transitions management – to create a variety of ways to understand and manage complexity. In the process, they transformed the meaning and significance of both science and management. They de-centered the privileged position scientists once enjoyed, and positioned them as one set of actors amongst others with distinct understandings of complex systems. Management of complex systems, in turn, requires collaboratively synthesizing these actors' diverse forms of knowledge and interests – a process that requires institutional change in order to develop trust amongst actors and facilitate reflexive and adaptive engagements with environmental phenomena.

However, in keeping with the genealogical spirit that animates this book, these efforts to transform thought and practice around a limited knowledge of complex worlds also have a unique history. As we suggested above, resilience proponents turned to the systems sciences and Herbert Simon's work on complexity and adaptive decision-making to inform their visions of complex social and ecological systems and adaptive management. The next chapter unpacks the links between Simon's thought and ecological resilience theory.

FURTHER READING

Brian Walker and David Salt's *Resilience Practice: Building Capacity to Absorb Disturbance and Maintain Function* offers theoretically nuanced and empirically descriptive discussions of how scholars and practitioners apply the insights of resilience theory to specific environmental management situations. It remains one of the most accessible descriptions of resilience theory in action. Foundational texts that laid the groundwork for resilience scholars' critical engagements with prevailing management paradigms include Kai Lee's *Compass and Gyroscope: Integrating Science and Politics for the Environment*, Elinor Ostrom's Nobel Prize-winning *Governing the Commons: The Evolution of Institutions for Collective Action* and C.S. Holling's *Adaptive Environmental Assessment and Management*.

5

RESILIENCE AS DESIGN

INTRODUCTION

Writing on the conceptual roots of panarchy, C.S. Holling and Geoff Sanderson (1996: 77–78) remark that,

> [Herbert] Simon was one of the first to describe the adaptive significance of hierarchical structures. He called them "hierarchies," but not in the sense of top-down sequence of authoritative control. Rather, semiautonomous levels are formed from the interactions among a set of variables that share similar speeds (and, we would add, geometries). Each level communicates a small set of information or quantity of material to the next higher (slower and coarser, in geometric terms) level. This "loose coupling" between levels allows a wide latitude of "experimentation"within levels, thereby increasing the speed of evolution.

This passage is repeated, almost verbatim, in several other essays written during the mid-1990s and early 2000s (Gunderson *et al.* 1995b: 518; Holling 2001: 392; Holling *et al.* 2002: 72). This was a period of conceptual fermentation when ecologists consolidated their definition of panarchy – and Herbert Simon is the key agitator. In each essay, this description of

Simon's definition leads into discussions of ecologists' sense of complexity. And by the early 2000s, this discussion culminates in a specification of panarchy. To be sure, ecologists slightly modify Simon's understanding of hierarchy to address ecological debates prevailing at the time (more on this below). But the larger point remains: this common trope signals Simon's singular influence on ecologists' understandings of complexity and panarchy – and thus his influence on how ecologists understand the topological transformations that make up resilience.

But this is far from the only prominent reference Simon's work receives. In his influential 1993 text on adaptive management, Kai Lee emphasizes how Simon's theory of bounded rationality underpins theories of adaptive environmental management. Lee (ibid.: 51–53) notes that Simon's understanding of rationality as bounded by limited human information-processing capabilities implies that humans pursue satisficing rather than optimizing decisions. That is, humans search for acceptable solutions to a given problem, not necessarily the best. For Lee (ibid.: 52), Simon's insight means that "inconsistency is normal" and that "large organizations can do things no single person can accomplish." More importantly, it *also* means that "learning in a world of bounded rationality is a costly, step-by-step search for better alternatives, in which local improvements may or may not benefit the whole" (ibid.: 53). This stepwise yet inconsistent search for acceptable solutions – one of the defining characteristics of adaptive management, as the previous chapter detailed – is nothing more nor less than "a strategy for using bounded rationality to learn rapidly," which Lee characterizes as "deliberate experimentation, which isolates part of complex reality, makes simple changes in it, and watches for results" (ibid.).

Simon's work *also* helped shape institutional analysis, which played an essential role in mainstreaming resilience approaches (see Chapter 4). In her Nobel Prize acceptance speech, Elinor Ostrom (2009: 409) credits Simon's work in public administration and behavioral economics with laying the groundwork for her influential Institutional Analysis and Development research framework. In particular, Simon's understanding of the hierarchical structure of decision-making provided a model for her multi-level "three worlds" of policy action and the different forms of rule-making decisions that occur on each level (Kiser and Ostrom 1982).

By their own accounts, Herbert Simon has clearly influenced many prominent resilience scholars. And yet, these conceptual roots remain largely under-acknowledged. This is perhaps unsurprising in an applied research field such as ecological resilience, which no longer concerns itself with foundational conceptual work. Instead, ecologists now pursue pragmatic solutions to pressing problems of complexity on the basis of conceptual foundations laid in the 1990s and early 2000s (see Walker *et al.* 2004; Folke 2006; Folke *et al.* 2007; Brand and Jax 2007 for reflections on the field's development). Why dwell on the intellectual grounds of panarchy, adaptive management, and institutional design if these concepts all "work" – that is, if they facilitate new critical interventions into social and environmental management practices? But many critical scholars have yet to engage with Simon's influence on ecological resilience, in part because they tend to emphasize formal affinities between resilience theory and neoliberal governance strategies (Walker and Cooper 2011). One exception is David Chandler (2013), who recognizes Simon's formative impact on institutional economics, and thus his influence on Friedrich Hayek's psychological writings and Anthony Giddens' "Third Way" governance strategies that underpin the soft paternalism prevalent in contemporary UK governance (see also Jones *et al.* 2013).

Chandler's genealogy offers a more careful unpacking of the relations between resilience-based governance strategies and neoliberal thinkers such as Hayek than many critical scholars provide.[1] In this chapter, I dwell on Simon's work in order to push this genealogical thread further. I will advance three main arguments. First, I will demonstrate how ecologists developed their sense of resilience within a thoroughly Simonian horizon of thought. Ecologists strategically picked up and modified key elements of Simon's thought – especially his concepts of hierarchy, bounded rationality and adaptive problem-solving – as they critically engaged with prevailing environmental science and management practices.

Second, foregrounding Simon's influence situates resilience within the wider genealogy of modern design. Simon's theories of complexity and bounded rationality were outgrowths of post-World War II cybernetics. Cyberneticians sought to develop a universally valid theory of change that could account for physical, chemical, biological, ecological, social and technological dynamics (DeLanda 1991; Kay 2000). They emphasized processes of control and communication, facilitated through information flow between an object and its environment. Many scholars have noted in passing the influence of cybernetics on resilience theory (Olsson *et al.* 2015; Folke 2006; Walker and Cooper 2011; Zebrowski 2016), but its wider implications still need to be teased out. Jean Baudrillard (1981) suggests that cybernetics emerged out of the consolidation of a *new sensorial regime* (Rancière 2004) in the first decades of the twentieth century. That is, cybernetics was conditioned by a new way of experiencing the world – a new way of sensing the world around us, orienting ourselves within this sensory environment, understanding how and why phenomena occur and acting within the world. Baudrillard identifies a "rational conception of environmental totality" as the precondition for cybernetics. With the term "environmental totality," he signals not the "nature" of modernist thought, an ontologically pure bio-physical realm cleansed of any infiltration from an equally pure "social" (cf. Latour 1993). Instead, it gestures towards the extension of rational abstraction to the whole of existence. This involves a "universal semantization of the environment in which everything becomes the object of a calculus of function and of signification." That is, we apprehend our surroundings as *objects with determined functions* (Baudrillard 1981: 185).[2] Baudrillard's sense of environmental totality thus signals an experience of the world as a functional system comprised of objects whose arrangements can be re/configured in particular ways according to the task (i.e., functional needs) at hand.

The links between Baudrillard's arguments and ecological theories of resilience, much like those between design and resilience, may not be readily apparent (see Chapter 1). But Baudrillard enables us to draw out a common thread across resilience theory, Simon's behavioral science, cybernetics, and design: each circulates within the same aesthetic regime of modern design. Despite considerable differences across these fields, they all share a common experience of world as environmental totality. In such a world, the truth of objects is less what they *are* than what they *can do*. The implications are significant: resilience theory reconfigures the study of human–environment relations around the aesthetics of modern design. It reduces the world to rational abstractions that can be functionally re/combined with one another. And it conceptualizes human–environment relations as a design problem that can be addressed through what I call, riffing off Simon (1996), a science of ecological artifice: exploring how to synthesize different social and environmental phenomena in order to create systems with certain properties – such as adaptive or transformative capacities that enhance social and ecological resilience.

Recognizing the aesthetic dimensions of resilience thus brings to light how resilience theory operates through a distinct relationship between truth and control. In a complex 'environment,' the truth of objects lies in their functionality. And their functionality, in turn, hinges on their relations with other objects. Control is no longer a matter of predictive knowledge, but is rather an ongoing process of learning about and managing feedback-driven relations between various entities that make up complex social and ecological systems. Rather than a will to truth that reduces reality to universally valid objective truths that can be folded into a calculative and predictive rationality, resilience theory mobilizes a *will to design* that strives to reduce reality to functional abstractions that can be re/combined with one another in order to solve contextually specific problems of social and ecological complexity.

This chapter unpacks the will to design that lies at the heart of resilience. Because many readers may be unfamiliar with Herbert Simon's work, it begins by reviewing two of his major theoretical contributions: his theory of bounded rationality and his understanding of complexity's hierarchical structure. It then explores these concepts' formative influence on resilience proponents' understandings of panarchy and adaptive management. After tracing these connections, I situate Simon's work within the broader context of cybernetics and the aesthetic regime of modern design. Here, Simon's arguments on the "sciences of the artificial" are particularly important, for they explicitly link his theories of hierarchy and bounded rationality with a modernist design aesthetic. The chapter concludes by examining how this aesthetic introduces a new relation between truth and control in both Simon's thought and ecological resilience theory.

DIAGRAMMING HERBERT SIMON

Herbert Simon does not slot easily into a single disciplinary identity. He received his PhD in political science from the University of Chicago in 1944, but went on to author field-defining works in economics, public administration, psychology, computer science and design studies. He was remarkably distinguished as well: among numerous honors, he received the Association for Computing Machinery's Turing Award in 1975 for his work on artificial intelligence, the American Psychological Association's Award for Outstanding Lifetime Contributions to Psychology in 1993, and the Nobel Prize in Economics in 1978 for his work on decision-making in organizations. Of course, he also engendered a fair share of controversy, particularly in fields that showed resistance to overtly cybernetic modes of explanation such as psychology or design (Mirowski 2002; see also Chapter 7). If we follow Norbert Wiener's definition of cybernetics as "the science of communication and control, whether in machines or living organisms" (Wiener 1964, in Bowker 1993: 113), then Simon was a quintessential "cyborg scientist" who sought to reconceptualize problem-solving as an adaptive, information-driven interaction between an inner and outer environment, common across all thinking creatures – human, animal or machine. Indeed, Simon never wavered from his controversial conviction that the human mind and the computer were topologically indistinct – for him, both followed the same satisficing problem-solving procedures.[3] This is perhaps the unifying theme of Simon's wide-ranging corpus: he is less a practitioner of disciplinary study than a cyborg social scientist who sought to uncover and analyze complex, hierarchical organizational structures wherever he looked – whether this was bureaucratic organizations, economic transactions, cognitive processes in the human mind, simulated cognitive processes in computer programs, or design – the creation of artifice (Sent 2000).

Indeed, Simon's thought traveled effectively across disciplinary boundaries because it was not tied to specific disciplinary debates, but instead offered an abstract diagram of cognitive behavior in complex situations. My use of the term *diagram*, as Chapter 1 discussed, draws on Deleuzian philosophy. Diagrams are real without being actual: they refer to an abstract series of relations between force relations that attempt to give meaning, structure and direction to an immanent field of indeterminate potentiality. In short, diagrams attempt to structure the possibilities for understanding and intervening in the world. Focusing on the diagrammatic qualities of Simon's thought directs our attention to the concepts he developed to understand complexity and human activity in this complex world. Concepts such as hierarchy, bounded rationality, near-decomposability, adaptation, synthesis and simulation form a series of relations that enable scholars in a number of disciplines to impose a certain kind of sense on an indeterminate world. In particular, resilience scholars mobilized this diagrammatic structure in their own practices of critique (see Chapter 4).

The genealogical approach I adopt here focuses on the series of continuities, disjunctions, and transformations that occurred as ecologists mobilized Simonian concepts to critically engage with prevailing thought and practice in the discipline. Before unearthing these designerly roots, I first detail Simon's theory of bounded rationality and the practices of critique and intervention his theory is founded upon.

BOUNDED RATIONALITY: RECALIBRATING RATIONALITY FOR A COMPLEX WORLD

As we saw in the previous chapter, resilience scholars in ecology developed their thought through a sustained critique of command and control. Simon's work similarly emerged through critical engagements with prevailing notions of rationality, decision-making and optimization in economics and first-order cybernetics research[4] in computer science (Mirowski 2002). While this chapter focuses almost exclusively on the former line of critique, the latter is useful here for illustrative purposes. Simon was a committed empiricist who maintained that theory should reflect the real world (even as he produced abstract academic theory). In his view, both neoclassical economics, associated with Walrasian equilibrium and Pareto optimization, and John von Neumann's work on automata offered elegant but empirically flawed decision-making theories. In the case of the former, neoclassical economics assumes that decision-makers are *homo economicus*, or rational "economic man" (and it is always a male subject). "Economic man" is a utility maximizer who, economists assume, possesses near-perfect knowledge of his environment, has a stable set of utility preferences, and is capable of determining the optimal course of action that will maximize the utility he gains from any transaction (Simon 1955: 99). In the case of the latter, von Neumann's theory of automata attempted to develop a computing system that would follow the same thought processes as humans and other thinking organisms – but he assumed that human cognition mirrored the pattern of formal logic. Formal logic provided him with a sequence of operations for the computing machine to follow; the challenge he faced was how to encode this deductive and mathematically oriented procedure into a language that could instruct computers how to independently (that is, autonomously) process information.

For Simon, these theories bore little resemblance to actual thought processes. He did not mince words while dismissing assumptions of *homo economicus*' calculative prowess: "there is a complete lack of evidence that, in actual human choice situations of any complexity, these computations can be, or in fact are, performed" (Simon 1955: 104). Theorizing decision-making instead required a more humble account of the economic subject. As he explained in his important 1955 paper:[5]

> the task is to replace the global rationality of economic man [*sic*] with a kind of rational behavior that is compatible with the access to information and the computational capacities that are actually possessed by organisms, including man [*sic*], in the kinds of environments in which such organisms exist.
>
> (Simon 1955: 99)

Simon thus set about rethinking rationality as a style of behavior inflected, but not determined, by the wider environment. He would later specify rationality as "the style of behavior that is appropriate to the achievement of given goals, within the limits imposed by given conditions and constraints" (Simon 1972b: 161). We will explore his functional conceptualization of environment below, for it provides an important link to modern design aesthetics. For now, what matters is that the environment "bounds" rational behavior. Its complexity exceeds totalizing objective comprehension. Decision-makers cannot access all information about complex environments, violating neoclassical assumptions of perfect knowledge. They also cannot process all data and information that is available to them, and thus cannot comprehend the complex web of interconnections in which they are caught up – which means they cannot anticipate and consider all possible outcomes, thus violating neoclassical assumptions that rationality necessarily entails optimization. Given these limitations, individuals do not optimize, but instead pursue a course of action they expect will deliver satisfactory results. Simon called this *satisficing* behavior: a kind of boundedly rational behavior that will produce "good enough" (rather than "the best") results.

Simon's understanding of satisficing behavior does not *reject* the principle of rationality as much as *recalibrate* it for a complex world where neoclassical assumptions of perfect knowledge, foresight and the rational application of deductive logic no longer hold. Two conceptual innovations facilitate this recalibration. First, he argues that boundedly rational decision-making processes adopt a *hierarchical* structure that disaggregates complexity into more manageable sub-units. Second, Simon characterizes decision-making as a stepwise, *adaptive* search for satisficing solutions, which allows solutions to inductively emerge through the decision-making process. These concepts provide an abstract rendering of complexity and human behavior that, according to Simon, could explain the dynamics of any complex system – social, biophysical or technical. Each innovation is considered in turn.

Disaggregating rationality: the hierarchical structure of decision-making

Simon's theory of bounded rationality does not take rationality for granted. Rather than assuming individuals are necessarily rational decision-makers, bounded rationality opens up the problem of *how* the individual engages with a complex environment. Bounded rationality signals a *process* of decision-making: setting goals, devising ways to meet these goals, and adjusting how these goals can be met as that environment kicks up new challenges along the way. In *Administrative Behavior*, first published in 1945, Simon decomposes rational behavior into hierarchically arranged levels (see especially Simon 1997: 106–111). As Holling (1986, 2001) would later recognize in terms we saw in Chapter 3, in Simon's formulation, hierarchy does not refer to relations of control or authority. Instead, hierarchy signals "a system that is composed of interrelated sub-systems, each of the latter being, in turn, hierarchic in structure until we reach some lowest level of elementary subsystem" (Simon 1962: 468). Hierarchy is thus a way to understand organized complexity in systems in which there is no subordinate relation among sub-systems. A hierarchical system comprises loosely coupled "levels" or "scales"; elements on each level share similar properties with one another (such as speed of movement or spatial extent) but have limited interaction with those on other levels.

Thus, in *Administrative Behavior*, Simon disaggregates organizational decision-making into functionally distinct kinds of decisions, spread across three hierarchically arranged levels: higher-level *substantive planning*, decisions on the organization's overarching values and goals; mid-level *procedural planning*, decisions on mechanisms that direct day-to-day activities towards the goals determined through substantive planning; and lower-level day-to-day decisions that make up everyday experience. For example, the overarching values of a municipal fire department include preventing fires and minimizing the damage of those that occur; mechanisms for achieving these goals include firefighting teams that respond to outbreaks and fire prevention units that work to mitigate risks, and daily activities include training activities, reviewing procedures and manuals, responding to emergencies, and so forth. In principle, effective administration integrates these hierarchical levels, so that all activities and decisions support the organization's overarching goals.

For Simon, this hierarchical structure facilitates engaging with complexity. Decisions made at higher levels reduce the complexity of decisions at lower levels. A fire department does not have to worry about policing or municipal taxation policy, because its overarching value directs its attention to the problem of fires.[6] Individuals within the fire department do not have to concern themselves with all aspects of fire management, because mid-level decisions provide procedures and organizational structures that allow some to focus on fire response and others on fire mitigation. Thus, decisions made to address specific problems found at higher levels reduce the field of problems decision-makers face at lower levels. Furthermore, those higher-level decisions tend to become ingrained as habit, memory, or institutional rules and procedures. They become "second nature," something the individual performs with limited conscious thought – which thus frees her attention to focus on more immediate problems located at lower levels of a decision hierarchy.[7]

Simon's understanding of hierarchy thus disaggregates rationality across functionally distinct levels. The task of administration is to integrate these levels in a way that enables individuals to achieve goals in a complex world despite their limited knowledge. If higher and lower levels align, then their decisions will produce satisficing outcomes even as they are enmeshed within complex environments.[8] And this is precisely Simon's unique definition of rationality given above: behavior that brings about desired goals within given conditions and constraints. But this sense of rationality does not entail utility maximization. Instead, it involves a stepwise and *adaptive search* for satisficing solutions that the hierarchical structure of decision-making facilitates. The next section explores this search process.

Distributing rationality: the adaptive search for satisficing solutions

In *Administrative Behavior*, Simon uses an example of civil engineering to contrast an inductive, stepwise and boundedly rational problem-solving approach with a deductive, logical and classically rational one. He suggests that an engineer designing a railroad to connect two cities through mountainous terrain will engage in the following procedure:

> After a preliminary examination of the topography, he [sic] selects two or three general routes that seem feasible. He then takes each of these routes as his [sic] new "end" – an intermediate end – and particularizes it further, using more detailed topographical maps. His [sic] thought process might be described as a series of hypothetical implications: "If I am to go from A to B, routes (1), (2), and (3) seem more feasible than others; if I am to follow route (1), plan (1a) seems preferable; if route (2), plan (2c); if route (3), plan (3a)" – and so on, until the most minute details of the design have been determined for

> two or three alternative plans. His [sic] final choice is among these detailed alternatives,
>
> (Simon 1997: 109)

Of course, this procedure follows a hierarchical structure: rather than considering all potential routes, the engineer first engages in a high-level choice (three general paths) that sharply narrows the field of possibilities. On the basis of more detailed study, he then develops more precise routes along each path and chooses the best-looking route among these options – a mid-level choice. He then conducts more detailed study among these alternatives in order to carry out the lowest-level decision on the exact route to build the railroad.

Simon then compares this hierarchically organized, boundedly rational decision-making process with an alternative:

> This process of thought may be contrasted with a single choice among *all* the possible routes. The latter method is the one dictated by logic, and is the only procedure that the decision arrived at is the best. On the other hand, this method requires that all the possible plans be worked out in full detail before any decision is reached. The practical impossibility of such a procedure is evident.
>
> (Ibid.)

This is the crux of the second limit to neoclassical models of rational behavior: the "practical impossibility" of knowing in advance the "full detail" of each and every choice. Faced with such a complex (yet banal) problem, the engineer draws on past experience, habit and training in order to inductively narrow down his choices and arrive at a satisfactory decision.

Simon characterizes this process of reflecting on progress towards a goal from a position of limited knowledge and foresight as an *adaptive* process. Historian of science and Simon biographer Hunter Crowther-Heyck (2005; see also Heyck 2008a, 2008b) suggests that Simon's theory of bounded rationality effectively substitutes *homo economicus* with *homo adaptivus*, or adaptive (wo)man. Adaptation was essential to Simon's recalibrated sense of rationality. He unequivocally equated the terms, writing with his frequent collaborator, the physicist Allen Newell, that,

> the behaviors commonly elicited when people (or animals) are placed in problem solving situations (and are motivated toward a goal) are

called *adaptive*, or *rational*. These terms denote that the behavior is appropriate to the goal in the light of the problem environment; it is the behavior demanded by the situation.

(Newell and Simon 1972: 53, emphasis in original)

Contemporary scholars of adaptation have long recognized the concept's roots in functionalist social science (Pelling 2003, 2010; Watts 2015). But Newell and Simon's definition is particularly useful because it foregrounds the importance of the "problem environment," and specifically the demands this environment places on the adaptive subject. They go on to suggest that, "if there is such a thing as behavior demanded by a situation, and if a subject exhibits it, then his [sic] behavior tells more about the task environment than about him ... if we put him in a different situation, he would behave differently" (Newell and Simon 1972: 53). The adaptive subject is thus a far cry from the sovereign Cartesian subject. The latter found certainty in the ontologically independent *cogito* – or the capacity for self-reflective thought expressed in Descartes' famous proof, *cogito, ergo sum*. A Cartesian subject is, in principle, the subject of a universally valid form of rational thought. It is always already detached from the world, absorbed in the mental activity of thought. In contrast, *homo adaptivus* is always embedded in the world. Its natural habitats are "problem situations" that demand responses. This is not a subject that can impose its will on the world; instead, a complex world always impresses upon the subject (Chandler 2013). And the adaptive subject can no longer hope to control the world; the best it can do is to hope it might adapt (Evans and Reid 2014).

Homo adaptivus is thus a topological figure: it straddles the threshold between an "external" environment that poses problems, and an "internal" consciousness that (in Newell and Simon's formulation) processes information from this environment and responds in an appropriate manner. Rather than modernist detachment from the world, which Descartes and later Kant both sought in the confines of consciousness, the adaptive subject is always at once inside and outside, never fully removed from its environment yet always reflecting on and adapting to the problems this environment poses. The process of decision-making is precisely the 'site' where she engages with a complex environment and adjusts her decisions – adapts – in order to realize a goal.

Synthesizing Simon's ontology

Together, Simon's specifications of hierarchy and adaptation enabled him to reconceptualize rationality in a complex world. Adaptation temporally redistributed rationality: rationality is dispersed across a stepwise and adaptive search for satisficing solutions, rather than expressed in a single optimization calculation. Hierarchy spatially redistributed rationality: the rational decision no longer took place on *homo economicus*'s perch atop the "head of a pin." Instead, Simon disaggregated the decision process across hierarchically structured levels that together comprised a cybernetic space: a relational space, functionally organized through feedback loops that integrated these otherwise disparate levels. These spatial and temporal reconfigurations are interrelated: as the example of the city engineer shows, movement from a higher to a lower level of decision-making often occurs through an adaptive search. Hierarchy thus enables Simon to understand processes of adaptation, and vice versa.

Moreover, hierarchy and adaptation are ontological concepts in Simon's thought. They describe what exists: a complex and hierarchically structured world populated with adaptive beings pursuing satisficing solutions to complex problems. Simon provides a functionalist explanation for this ontology: the hierarchical organization of complexity allows for change, experimentation, and evolution to occur on the level of one subsystem, insulated from the rest of the system – a point resilience scholars enthusiastically picked up in their efforts to understand complex systemic dynamics (Holling 1986).

This functional efficiency results from what Simon calls *near-decomposability*, or the tendency for elements within one hierarchic level to interact with others on the same level, and remain more or less functionally independent of other scales. A brief and oft-quoted parable will illustrate this point. Simon (1962) asks us to imagine two watchmakers, each attempting to build watches that consist of 1,000 parts. But there is a catch: if something interrupts them as they are in the process of assembling the watches, their progress is lost. The watch falls apart and they have to start over. The first watchmaker attempts to assemble the complete watch from start to finish. Needless to say, he (and again, they are always men) is not able to complete many watches; no matter his individual skill or speed, each time he gets interrupted he has to go back to square one. The second watchmaker uses a process that allows him to compartmentalize his craft. Rather than attempt to build an entire watch, he assembles some of the parts into stable components, analogous to nearly decomposable sub-systems in Simon's hierarchy. Thus, when he is interrupted, he only loses progress on that particular component, and can adapt to the disturbance and regain his place relatively quickly.[9] Through quantitative analysis, Simon shows that the second watchmaker will rapidly out-produce the first.

The diagrammatic qualities of Simon's thought are evident in this parable. Here, hierarchy indicates an abstract arrangement between watch parts, the craftsman, and the disruptive environment. This diagrammatic set of relations allows complex organization to emerge – in this case, the completed watch. For Simon, hierarchy is an abstract universal, a form that exists independent of content: "complexity frequently takes the form of hierarchy, and ... hierarchic systems have some common properties that are independent of their specific content" (Simon 1962: 468). For historian of science Esther-Mirjam Sent (2001), Simon's ontological sense of hierarchy is a unifying theme across his half-century of varied disciplinary pursuits. As she details, Simon saw hierarchies wherever he looked – whether this was in human organizations, economic transactions, cognitive processes, computing systems, computer languages, the design and function of machines, the practice of urban planning, and so on. Common hierarchical properties can be found in a very broad class of complex systems – from the physical, chemical, and biological to the social and the technical (Simon 1962) – because hierarchies facilitate the emergence and stabilization of complexity itself. Thus, complexity is always evidence of ontologically prior hierarchical structures that enabled complexity to evolve out of relatively simple starting conditions.

Simon's abstract diagram of complex and adaptive systems infused thought on complexity and change in a number of fields, including ecology. His critique of deductive decision-making models also provided a blueprint of sorts for resilience scholars' own critical engagement with dominant environmental science and management practices, to which we now turn.

DIAGRAMMING RESILIENCE

Ecologists widely recognize that resilience theory emerged out of the earlier development of systems ecology in the 1930s and 1940s (Berkes and Folke 2003; Folke 2006; Olsson *et al.* 2015). Systems theory provides the basic tenets of resilience thinking: connectedness, intra- and cross-scale feedback, self-organization, and a functional understanding of system components and their dynamics (that is, each element in a system derives its identity and meaning from its functional relation with other elements). However, resilience scholars differentiated themselves from other forms of systems analysis in ecology (e.g., Odum 1969; Allen and Star 1982) by explicitly drawing on Simonian understandings of connectedness, feedback, and functionalism. This influence is particularly evident in their understandings of panarchy and adaptive management. While resilience research today has largely black-boxed these concepts (that is, treats them as more or less finished and unproblematic 'objects' of thought that can work as the building blocks of further research), unpacking them shows how formative work in resilience picked up, modified, and deployed Simon's ontology. We first explore the Simonian roots of panarchy.

Transforming hierarchy into panarchy

Holling and colleagues note that a reprinted (and slightly revised) version of Simon's "seminal" 1962 article, "The Architecture of Complexity," inspired their theory of panarchy (specifically, Simon 1972a: see Holling and Sanderson 1996; Holling 2001; Holling *et al.* 2002). Simon's arguments enabled resilience scholars to advance at least four critiques against what they saw as problematic limitations in ecology.

First, Simon's understanding of hierarchy provides an important corrective for much ecological research on system dynamics and evolution. In his important 1986 article, Holling (1986: 73–74) equates ecological understandings of resilience with a distinct vision of nature: what he calls "nature evolving." This recognizes that nature is not static or machinic, but rather undergoes evolutionary change. For Holling (ibid.: 74), "the study of evolution requires not only concepts of function but also concepts of organization – of the way elements are connected within subsystems and the way subsystems are embedded within larger systems." As Chapter 3 discussed, in this article, Holling worked through the problem of how to connect local ecosystem dynamics with broader global environmental processes. Simon's understanding of hierarchy is crucial to linking the local and the global – and thus to ecologists' ability to intervene in ongoing debates over global environmental change and sustainable development. Holling (ibid.: 75) suggests that such concepts of organizational structure and cross-scalar dynamics are "connected in turn to hierarchy theory on the one hand, and the stability and resilience concepts ... on the other” and that they are “starting to provide the framework required for comprehending organizational evolution … such developments are an essential part of any effort to understand, guide or adapt to global change.” Later in the article, Holling (ibid.: 77) specifies this framework in familiar Simonian themes. Different scales represent ecosystems, defined as “communities of organisms in which internal interactions between the organisms determine behavior more than do external biological events.” Global events do impact ecosystems, but their effects “are mediated through strong biological interactions within the ecosystems. It is through such external links that ecosystems become part of the global system.” Simon’s vision of hierarchy thus provides ecologists with their own ontology of hierarchically organized complex ecosystems: tightly coupled sub-systems loosely connected to sub-systems at higher and lower scales.

Second, Simon’s sense of hierarchy provided a more tractable understanding of complex, ecosystemic interconnection. For Holling, the common ecological assumption that “everything is connected to everything else” is, “simply not true … the persistence of a species would be precarious indeed if its fate depended on every other species in a system … [instead,] ecosystems exhibit patterns of connections resulting in subassemblies that are tightly connected with themselves but loosely connected to others” (Holling 1978: 27). These “subassemblies” are Simonian lower-level sub-systems whose elements exhibit strong interconnections amongst each other but weak connections with higher and lower levels. Just as Simon’s sense of hierarchy enabled him to spatially and temporally redistribute rationality across different levels – and thus analytically reduce complexity – so too do ecologists deploy this sense of hierarchy to reduce ecosystemic complexity to manageable analytical levels. Holling argues in his 1986 essay that,

> ecosystems have a hierarchical structure, and for this reason it has been possible to capture the essential discontinuous behaviors with three sets of variables operating at different speeds. Other species and variables are dramatically affected by that structure and the resultant behavior, but do not directly contribute to it. Hence the relevant measures of species diversity, which is one measure of complexity, should not involve all species, but only those contributing to the physical structure and dynamics.

Readers versed in ecological literature will recognize here a generic definition of keystone species: species whose survival and persistence play a vital role in an ecosystem's functional stability. Indeed, elsewhere in the essay (Holling 1986: 74), Holling references ecological and biological research on food webs (which developed the concept of keystone species – see Paine 1966, 1969) as an example of applying concepts of hierarchical organization to the study of complex ecosystems. But Holling calibrates this concept in terms of the hierarchical organization of ecosystems. Keystone species compose a higher-level sub-system that structures the ecosystem as a whole. This means that species diversity, measured in terms of *total* number of species (and number of members of each species) is inconsequential: even if ecologists could somehow successfully catalogue all species within a system, the system's hierarchical structure, and the near decomposability of levels within this structure, mean that system dynamics only reflect the relations between a few keystone species. Moreover, such totalizing calculation is a methodologically impossible task (just as an engineer cannot practically consider all possible routes for a new railroad), and the pursuit of such totalizing vision is foolhardy: as Brian Walker and colleagues suggest, more complex models will mask system dynamics because "generally humans can only understand low-dimensional systems and because, empirically, it appears that only a few variables are ever dominant in observed system dynamics" (Walker *et al.* 2004: unpag.). This is Simon's theory of bounded rationality applied to the analysis of ecosystems: human cognitive and computational limits, coupled with complex empirical reality, render total knowledge of ecosystems impractical and pragmatically unnecessary.

Third, Simonian hierarchy has important methodological implications. The prevailing wisdom among ecologists and environmental managers at the time was to catalogue all species present within a system – what Holling (1978: 3) referred to as "comprehensive 'state of the system' surveys ... [that are] often extremely expensive yet produce nothing but masses of uninterpreted and descriptive data." But Simon's understanding of near-decomposability, translated into biological and ecological debates on keystone species, implies that the precise identity of species matters less than the functional role they play in sustaining systemic persistence. This has profound consequences for environmental management: "the conclusion for environmental assessment is that even qualitative measurements of structure are more important than measurements of numbers of every organism possible. The structure depends on who is connected to whom and how" (ibid.; see also Peterson 2000). What matters more than precise knowledge of all species, as Kai Lee (1993) demonstrates in the case of salmon conservation in the US Pacific Northwest, is to have a general model of the system in question – that is, a broad understanding of the qualitative relations between elements that structure systemic dynamics. Holling (1978: 35) drives this point home: "since everything is not intimately connected to everything else, there is no need to measure everything. There is a need, however, to determine the significant connections." As the previous chapter argued in terms of adaptive management techniques, simulations enable resilience scholars to test and reflexively adapt their understanding of these connections – something we will examine further below.

Fourth, ecologists also turned to Simon to correct what they saw as a fatal flaw in dominant disciplinary renderings of hierarchy. For Holling and others, early efforts to fold Simon into ecological theory problematically allowed conventional understandings of hierarchy as an authoritative relation to color their work. In the process, they lost "the dynamic and adaptive nature of such nested [hierarchical] structures" (e.g., Holling and Sanderson 1996: 78). Against this reading, Holling and colleagues articulated Simon's understanding of hierarchy with Holling's (1986) previous work on the adaptive cycle. They recognized that "Simon's key arguments are that each of the levels of a dynamic hierarchy serves two functions: one is to conserve and stabilize conditions for faster and smaller levels; the other is to generate and test innovations by experiments occurring within a level." For ecologists, "it is this latter, dynamic function we call an 'adaptive cycle'" (Holling 2001: 393) – but as we saw above, this is the adaptive process Simon details in his parable of the watchmakers. Holling and colleagues thus find in Simon a skeletal outline of resilience theory itself. The adaptive cycle, facilitated through hierarchical organization of complex systems, "explicitly introduces mutations and rearrangements as a periodic process within each hierarchical level in a way that partially insulates the resulting experiments from destroying the integrity of the whole structure" (Holling and Sanderson 1996: 78). And this understanding of adaptive cycle-*qua*-Simonian hierarchy overcomes the static vision of hierarchy ecologists had mobilized up to that point: "the adaptive cycle ... transforms hierarchies from fixed static structures to dynamic, adaptive entities whose levels are sensitive to small disturbances at the transition from growth to collapse (the omega phase) and the transition from reorganization to rapid growth (the alpha phase)" (Holling 2001: 396–397).

Simon's understanding of hierarchy thus provided resilience scholars with a foundation for their conceptual innovations and critical interventions during the 1970s and 1980s. Of course, ecologists also engaged with a variety of other influential concepts during this time, such as Schumpeter's theory of creative destruction (Cavanagh 2016), Dewey's pragmatic philosophy (Schmidt 2015), functionalist approaches to adaptation in anthropology and human ecology (Watts 2015; Pelling 2010) and biological understandings of life as information (Evans and Reid 2014), to name a few. But Simon's thought formed the horizon in which these other concepts gained meaning and significance for debates in ecology. For example, Simonian hierarchy enabled ecologists to visualize a series of nested adaptive cycles, one phase of which (Ω) involved Schumpeterian processes of creative destruction (Holling 1986, 2001). Complex ecosystems' functional adaptations played out within a Simonian sub-system largely insulated from dynamics at other levels. And the exchange of information that drove cybernetic processes of control and communication occurred within and across Simonian sub-systems whose elements were tightly coupled to each other and loosely bound to larger and smaller subsystems. Simon thus provided a conceptual apparatus that enabled some ecologists to reconfigure existing understandings of ecosystems, and existing problems in ecological research, around an ontology of hierarchically structured complexity. This also enabled resilience scholars to reconceptualize problems of environmental management and sustainable development around a Simonian understanding of adaptation as a stepwise search for satisficing alternatives, to which we now turn.

Transforming bounded rationality into adaptive management

As Chapter 3 demonstrated, ecological resilience emerged through critical engagements with prevailing approaches to sustainable development. Among other arguments, resilience scholars asserted that achieving sustainability in a complex world required adopting adaptive management strategies. Proponents drew heavily on Simon's understandings of bounded rationality and adaptation to formulate this alternative strategy. Kai Lee's influential 1993 study of environmental management in the US Pacific Northwest frames the challenge in Simonian terms. For Lee, adaptive management enables scholars to address the disjunction between *present* unstable and unsustainable social and ecological relations, and a sustainable social and ecological *future*. This is a quintessentially Simonian problem. Environmental managers cannot deduce the optimal strategy to achieve sustainability. Ecosystems are too large and complex; their management involves a fragmented landscape of overlapping, multi-scale and multi-agency jurisdictions, and there are multiple competing interest groups, often with incompatible and contradictory value systems. Facing such complexity, managing for sustainability becomes a stepwise and adaptive search for pragmatically feasible (i.e., satisficing) solutions that can balance these competing interests.

However, Lee's engagement with pragmatic philosophy – especially John Dewey's thought – leads him to emphasize *learning* as the essential process that mediates between an unsustainable present and sustainable future: "in the case of large ecosystems, pragmatism is a prime virtue: to learn what we can, and to recognize its limits" (ibid.: 11). Lee's blending of Simon and Dewey positions Simon's understanding of bounded rationality as the theoretical basis for social learning. If "social learning itself depends on the concept of bounded rationality" (ibid.: 53), then "the most important implication is that learning in a world of bounded rationality is a costly, step-by-step search for better alternatives, in which local improvements may or may not benefit the whole." Lee's Simonian characterization of learning as a step-by-step search repeats a common trope among early adaptive management proponents, whether this is Walters and Hilborn's (1978: 183) characterization of adaptive management as a "problem of sequential experimental design" or Holling's (1978: 1) description of a "sequence of steps." But Lee's emphasis on learning focuses attention on the gap between the current and desired situations – a gap that can be filled through adaptive management initiatives.

Carl Folke and colleagues, writing in their influential manuscript on institutional fit (Folke *et al.* 2007), share Lee's assertion that bounded rationality provides the basis for both human institutions and institutional learning. They define the latter as "collaborative and collective learning that becomes embedded in institutions" (ibid.: unpag.).[10] While their discussion of bounded rationality references neoclassical economists – whose appropriations of Simon's concept whitewash many of his more radical critiques (Mirowski 2002; Sent 2005) – they nonetheless suggest that an analytical focus on learning, founded on bounded rationality, has been essential for new research into adaptive co-management and adaptive governance. Indeed, a focus on learning enables resilience scholars to fold error and surprise into scientific research and management strategies. Lee asserts that adaptive management emerged out of the "bias that understanding how natural systems respond to human disturbance is essential for 'living with the unexpected'" (Lee 1993: 54, quoting Holling 1978).

In a complex world, knowledge is always partial and rationality is always bounded, and the "unexpected" is an unavoidable part of life. Management must learn to deal with surprise, rather than pathologically attempting to command and control nature through prediction and eliminating error (Holling and Meffe 1996). In this context, Lee (1993: 56) suggests that, "because human understanding of nature is imperfect [i.e., bounded], human interactions with nature should be experimental. Adaptive management applies the concept of experimentation to the design and implementation of natural resource and environmental policies" (Lee 1993: 53). However, as the previous chapter detailed, experimentation is not geared towards testing and refuting hypotheses. Instead, experiments bring about surprises that can enable learning: "if resource management is recognized to be inherently uncertain, the surprises become opportunities to learn, rather than failure to predict" (ibid.: 56).

Folke and colleagues echo Lee when they suggest that collective, institutional learning requires surprise and crisis, which results from the bounded nature of knowledge. Critical scholars have also emphasized this point in a variety of ways. For example, David Chandler (2014a) suggests that resilience transvalues evil – that is, rather than trying to prevent or avoid "bad" occurrences, resilience welcomes loss, failure and error as preconditions for growth and change. Similarly, Brad Evans and Julian Reid (2014) show how resilience borrows from the life sciences the assumption that life requires exposure to danger in order to develop adaptive capacities. However, we can see here how this transvaluation of evil, error and crisis results from a Simonian sense of bounded rationality. Lee's work is exemplary on this point. He establishes an indexical chain that links together learning, sustainable development, experimentation, and bounded rationality: *sustainable development* can be achieved through *learning*; *learning* occurs through experiencing and reflecting on *surprise*; *surprise* is generated through *experimentation*; and *experimentation* is necessary because our *bounded rationality* inhibits our understanding of, and ability to control, complex reality. Or, as he puts it, "without experimentation, reliable knowledge accumulates slowly, and without reliable knowledge there can be neither social learning nor sustainable development" (Lee 1993: 54).

Synthesizing Simon's influence on ecological resilience

Unpacking the black boxes of panarchy and adaptive management demonstrates how Simon's thought sits at the heart of ecological resilience. His sense of hierarchy offers a vision of complex interconnection and emergent, self-organized topological change – the raw ingredients of the "very different view of the world" that Holling (1973: 15) suggested could transform management and science. His theory of bounded rationality enframes adaptive behavior as an ongoing, stepwise search for satisficing solutions – the raw ingredients of adaptive management and adaptive governance. These concepts thus allowed resilience scholars to critically reconceptualize the problems complexity posed to achieving sustainable development.

Recognizing Simon's influence suggests that the genealogy of resilience lies along a Simonian path. While resilience scholars' critiques of command and control undoubtedly share many commonalities with neoliberal economists' critiques of centralized economic planning (Walker and Cooper 2011), resilience theory exceeds these formal affinities with neoliberalism. This is a point that David Chandler and Jessica Schmidt have recently begun exploring: Chandler (2013, 2016b) indicates that the roots of resilience thinking lie in new institutional economics of Simon and John Commons; and Schmidt (2015) convincingly argues that resilience theory draws heavily on John Dewey. Both authors suggest that resilience is one way of envisioning how the inner world of the subject shapes, and is shaped by, the exterior environment – and that it is irreducible to neoliberal formulations of this topological field found in the work of Friedrich Hayek. This is an important argument, for it compels critical analysts to explore the ethical and political dimensions of resilience without recourse to the crutch of neoliberalism and the body of critique that has developed around this concept (Anderson 2015; Simon and Randalls 2016). However, Chandler and Schmidt both tend to pass over a key area: the threshold between inner and outer environment that Simon (1996), writing in his important *The Sciences of the Artificial*, calls artifice. Indeed, Schmidt (2013) explicitly argues that resilience is founded on the rejection of artifice. Here, she has in mind a modernist sense of artifice as the direct product of human will and the binary opposite of nature. Resilience scholars, like pragmatists and neoliberal thinkers, certainty do reject *this* vision of artifice. But other formulations of artifice are possible. Indeed, Simon's *Sciences of the Artificial* offers an extended study of artifice as a product of distinct style of thought and action, which he calls design. Simon proposed his understanding of design as a radical alternative to conventional social-scientific analyses premised on analytical methods and assumptions of predictive certainty (Mirowski 2002).

Just as he created a novel understanding of rationality to account for the challenges of complexity, he similarly created a novel understanding of social science compatible with a complex world that outruns predictive analysis. Design is an art of the threshold – a synthetic mode of thought that seeks to align interior and exterior worlds in the absence of transcendental determinants. And yet, critical scholarship has thus far mostly *avoided* exploring how this unique style of thought links to resilience theory. The remainder of this chapter begins unpacking these connections, starting with Simon's sense of design.

SIMON'S DESIGNER SCIENCE

In 1968, Simon gave a series of lectures at MIT that consolidated his wide-ranging oeuvre and articulated what he called "sciences of the artificial." In these lectures, published in 1969 as *The Sciences of the Artificial* (later expanded in a revised second edition, containing his 1981 Gaither lectures at Berkeley, and an updated third edition in 1996), Simon makes the case for a novel field of scientific inquiry, distinct from the natural sciences, that focuses on the process of adaptive problem-solving in a complex world. Simon's sciences of the artificial concern themselves with the process of producing artifacts – a process, he suggests, that the natural sciences are wholly unsuited to understand (Mirowski 2002: 471).

Simon (1996: 6) defines *artifact* in topological terms as "a meeting point – an 'interface' in today's terms – between an 'inner' environment, the substance and organization of the artifact itself, and an 'outer' environment, the surrounding in which it operates." While we might typically recognize an artifact as anything engineered by humans to serve a particular function and meet a specific goal (ibid.: 4–5), for Simon this interface provides a model for *any* entity – human or non-human organism, institution, machine – that attempts to align interior and exterior environments. For example, a well-designed wing will allow an airplane to fly, a well-designed clock will accurately measure and display time, a well-designed urban development plan will manage a society's urban growth. In terms of resilience, a well-designed environmental management institution will allow society to adapt to a complex environment. Thus, artifacts are not merely manufactured entities; more essentially, they embody and express goal-oriented adaptive behavior: "artificial things are often discussed, particularly when they are being designed, in terms of imperatives as well as descriptives" (ibid.). Because this imperative centers around aligning an interior with an exterior, artifacts are functional: "if the inner environment is appropriate to the outer environment, or vice versa, the artifact will serve its intended purpose" (ibid.: 6).

Simon's functional understanding of artifact focuses more on what artifacts *do* – and *can* do – than what they *are*. Studying artifacts does not involve steady-state descriptions; rather, it involves thinking in terms of functions, goals, and adaptation. And it directs attention to processes of setting goals and designing artifacts that can realize these goals. Simon's sciences of the artificial are, essentially, a science of *design*: the process of aligning (or adapting) interior to exterior in the absence of external, transcendent determinants of meaning and value. In Simon's (1996: 114) terms, design is "concerned with how things ought to be, with devising artifacts to attain goals." Design for Simon is a creative act; it involves the discovery and elaboration of alternatives. This is an iterative process of building alternatives and selecting satisficing solutions from amongst these alternatives (Simon 1972b). The contrast with classical decision theory is important here: if classical decision theory assumes an individual chooses from a predetermined set of alternatives, Simon's empirical sense of design recognizes that, in practice, individuals create alternatives out of nothing more than the pre-existing elements that comprise the interior and exterior environments.

This creative process operates through a distinct, synthetic style of thought (Cross 1982). Rather than the analytical thinking that dominates the natural sciences, and which attempts to explain and understand what something is or how something works, design revolves around synthesis. Synthetic reasoning is a future-oriented, topological mode of thought attuned to the *potential* ways elements can be re/combined with one another, and the pragmatic *possibilities* that result from these re/combinations. In other words, synthetic thought is concerned with how distinct entities might be brought together in order to meet a particular goal – whether this goal is to design a path for a road through a mountain range, a system for managing an organization's logistics, a cure for an ailment, a musical composition, a compelling narrative, or an article of clothing. Simon's (1996: 119) oft-quoted claim that "everyone designs who devises a course of action for changing existing situations into preferred ones" signals this common style of thought: each individual – planner, administrator, doctor, musician, writer, fashion designer – engages in the same style of synthetic reasoning.

Simon's arguments in *The Sciences of the Artificial* contributed to the formation of design studies, a distinct discipline that began emerging in the

1950s but consolidated in the 1960s, thanks in part to Simon's intervention (Cross 1982). As with Simon's work in other fields, his understanding of design is singular and reflects the distinct way he uses – synthesizes! – concepts such as bounded rationality and the hierarchical organization of complexity to address problems in specific disciplinary debates (Huppatz 2015). As we will see in Chapter 7, design studies scholars have branched off from Simon in a variety of ways, many of which are quite critical of Simon's scientistic approach. For now though, his understanding of design as a distinct way of understanding adaptive behavior in a complex world helps us situate resilience thinking within a particular historical context: the emergence of cybernetic thought that shaped Simon's own understandings of complexity, adaptation and design. Indeed, Simon's thought can be understood as a social scientific expression of wider cultural and economic changes that conditioned the emergence of cybernetics, to which we now turn.

SIMON AS CYBORG SCIENTIST

Simon's sciences of the artificial were never intended to *replace* the natural sciences. Instead, he brackets the natural sciences in order to create a space for his unique slant on social science (Mirowski 2002). He recognized that the former's analytical techniques were appropriate for producing objective and descriptive accounts of certain physical, biological and chemical phenomena – those elements that make up the outer and inner environments (Simon 1996). However, they are less appropriate for understanding the future-oriented, pragmatic and goal-oriented processes that synthesize interior and exterior. Accordingly, his mathematically oriented vision of the social sciences did not attempt to mimic physics or thermodynamics, a path neoclassical economists after Pareto pursued (Crowther-Heyck 2005). Rather, he drew inspiration from the interdisciplinary field of cybernetics research. The cybernetic study of communication and control in complex systems shaped Simon's vision of humans, non-human organisms, social organizations, and machines as information processors.

As genealogies of cybernetics have demonstrated, cybernetics research emerged in response to a certain kind of problem planners and engineers in the private sector, government, and military began facing in the 1920s and 1930s. On the surface, these actors faced nominally distinct challenges, such as building and maintaining telephone communication systems or large-scale power generation and distribution systems (Heyck

2008a), maintaining public order and national economic productivity in the face of increasingly militant labor uprisings (Duffield 2011), planning defense strategies against air warfare attacks (Collier and Lakoff 2008; Dillon and Reid 2009), or organizing communication and decision-making processes in the "fog and friction" of the battlefield (DeLanda 1991). However, at an abstract level, these all boiled down to the same basic problems of interconnection and emergence. Interdependence allowed isolated actions in one area to cascade across a system, leading to disruption and, potentially, catastrophic breakdown. Moreover, these events were non-linear; they could not be predicted in advance. Complexity thus rendered ineffective modernist techniques of security premised on spatial separation and predictive control.

During the 1930s and 1940s, the field of cybernetics gradually consolidated in response to these failures. Key here was cyberneticians' rendering of complexity in machinic terms. In their formulation, a machine was any device that transformed an input into an output, with a given amount of error, according to certain rules (Heyck 2008a). Arturo Rosenblueth, writing with Norbert Wiener and Julian Bigelow in a foundational essay, defines cybernetics (in their terminology, "behavioral science"), in telling terms:

> Given any object, relatively abstracted from its surroundings for study, the behavioristic approach consists in the examination of the output of the object and of the relations of this output to the input. By output is meant any change produced in the surroundings by the object. By input, conversely, is meant any event external to the object that modifies this object in any manner.
>
> (Rosenblueth *et al.* 1943: 18)

Rosenblueth and colleagues' definition locates the object firmly within a web of interconnections: it can impact its environment (output) but also be impacted by this environment (input). These connections to the wider environment are essential to the new "behavioristic" approach to scientific study they map out: in contrast to functionalist analyses, which they suggest emphasize an object's intrinsic systemic and structural order, their approach concerns itself with the *relation* between the object and its surroundings. And this relation depends on the circulation of information between object and environment. Information flow allows the object to adapt its behavior to its environment in a way that achieves a goal, which they define as "a final condition in which the behaving object reaches a definite correlation in time or in space with respect to another object or event" (ibid.).

The servomechanism is a quintessential example of this kind of goal-oriented, feedback-driven behavior. This term refers to an automated device that monitors the relation between interior and exterior, identifies and adjusts for error. A classic case is the thermostat, which measures the temperature of an external environment, compares it to a goal (the thermostat setting), releases energy to power an air conditioning device that will raise or lower the temperature (according to the direction of the error), and then shuts off the energy flow once the external environment's temperature equals the setting (Simon 1957). Key for our purposes here, servomechanism theory introduced new possibilities of control: feedback mechanisms – that is, the flow of information – can control purposive behavior in electronic machines, often with much more speed and precision than human activity (Heyck 2008a). The implications were profound: in a complex and interconnected environment that exceeded predictive or mechanical (i.e., linear) control, planners and engineers could still direct system performance through feedback. The challenge became designing devices that could identify relevant information signals within the surrounding environment, exclude irrelevant and unexpected noise, and adjust system performance in a way that would correlate performance with a pre-determined goal.

However, cybernetics did not congeal until Norbert Wiener, a physicist at MIT, began attempting to apply servomechanism theory to predict human behavior. Working in the nadir of World War II, as the British were besieged by Nazi air raids, Wiener strove to develop an "antiaircraft predictor" that would observe an enemy pilot's erratic flight path through shrapnel, smoke and explosions, anticipate his future position, and launch an antiaircraft shell to that location to destroy the plane. While Wiener's efforts met with middling success, he became convinced that servomechanism theory could be used to understand adaptive human behavior (Galison 1994). He first elaborated this argument in the above-referenced short essay co-authored with Rosenblueth and Bigelow (Rosenblueth *et al.* 1943), and developed these arguments in a number of influential publications over the next twenty years. In large part through Wiener's influence, cybernetics solidified as a trans-disciplinary endeavor that promised to offer a universally valid understanding of systemic behavior. Whether the system involved psychological processes, human interactions, technological devices, social organizations, or human–machine assemblages, each could be analyzed in terms of servomechanism theory – that is, in terms of information flow and adaptive, goal-oriented behavior.[11]

Simon both drew on and critiqued Wiener's formulation of cybernetics. At times, he approvingly cites Wiener, who he suggested "has argued persuasively that the servomechanism model may be a useful model for describing physiological, psychological and sociological adaptive systems" (Simon 1969: 399). And he mathematically demonstrated that servomechanism theory could be used to design an efficient logistics system in large organizations (Simon 1957). However, he also acknowledged significant limitations. While servomechanism theory might provide an effective understanding of closed-loop systems (such as a thermostat, organizational logistics, or torpedo equipped with a target-seeking mechanism), it provided a back door into optimization when applied to human decision-making processes. Specifically, it assumes a system can be designed in a way to reliably perform at an externally determined target level. But for Simon, as always, the social world is far more empirically complex. In a discussion on social science modeling methodology, Simon (1969) suggests that goals may change, and that they may be influenced by external, "intervening" forces that are not a part of the system's normal operation. Simon's argument here is more in line with second-order cybernetics (akin to Chandler's (2014a) *general complexity*), which asserts that the observer is also part of the complex system. This means that the goals themselves may change as the observer receives information about the environment. In contrast to the closed-loop system of servomechanism theory, which assumes a single equilibrium state, Simon suggests that open systems can be characterized by shifts between different equilibrium points, and that external conditions will spark this shift – precisely the vision of multiple equilibrium systems Holling (1973) would soon articulate within ecology.

Despite his deviance from Wiener's servomechanism theory,[12] Simon nonetheless shares Wiener's cybernetic inclination to reduce life itself to context-free information exchange. The philosopher of science Georges Canguilhem (1994: 317), discussing the impact of cybernetics on the life sciences, explains that after the cybernetic revolution, "in order to understand living things one needs a non-metric theory of space, a science of order, a topology: one needs a nonnumerical calculus, a combinatorics, a statistical machinery." Simon (1996: 5) echoes Canguilhem's vision of an informatics-oriented life science when he characterizes the object of a science of the artificial – the adaptive process of aligning interior and exterior – as an "information-centered process of adaptation to a world that can only be known through the models we construct to represent it." This epistemology of models – what Simon (ibid.: 13) describes in terms of *simulation* – hinges on the cybernetic assumption that adaptive behavior can ultimately be reduced to abstract information exchange between an interior and exterior, facilitated by an artifact that straddles the threshold. Simon (ibid.: 13, emphasis added) suggests that,

> the artificial object imitates the real by turning the same face to the outer system, by adapting, relative to the same goals, to comparable ranges of external tasks. Imitation is possible because distinct physical systems can be organized to exhibit nearly identical behavior. *The damped spring and the damped circuit obey the same second-order linear differential equations*, hence we may use either one to imitate the other.

Simulating complex systems through modeling produces tractable knowledge because, for Simon and other cyberneticians, the world is already given to modeling. Machines, humans, non-human organisms – each is enmeshed within processes of information exchange, and mathematical equations can describe these processes. But this requires a different use of mathematics in line with Canguilhem's description of a nonnumerical calculus: "numbers are not the name of this game but rather representation structures that permit functional reasoning, however qualitative it may be." (ibid.: 146). Models thus facilitate the development of a specific kind of knowledge – the functional reasoning that discloses what objects do and can do as they are synthesized with other objects.

As numerous historians of science emphasize, Simon's ability to conceptualize an epistemology of simulation hinged on his use of, and access to computers.[13] In Simon's hands, the computer became a tool for understanding human behavior, because each engaged in the same process of symbol processing. He suggests that the computer is a member of "an important family of artifacts called symbol systems … another important member of the family (some of us think, anthropomorphically, it is the *most* important) is the human mind and brain" (Simon 1996: 21, emphasis in original). He defines symbol systems as "machines" that produce "symbol structures," or representations (ideas) of the external environments, which enable the machine to reflect on this environment, devise

a course of action, and follow this course of action to realize a goal. If we follow Simon's equation of rationality with adaptation, as above, then both humans and machines exhibit rational behavior. But while human information processing largely remained a mystery, its cybernetic equivalence with computing machines meant that computers could simulate those processes and thus produce knowledge on human problem-solving. As he notes in designerly terms:

> the relation of [computer] program to environment opened up an exceedingly important role for computer simulation as a tool for achieving a deeper understanding of human behavior. For if it is the organization of components, and not their physical properties, that largely determines behavior, and if computers are organized somewhat in the image of man, then the computer becomes an obvious device for exploring the consequences of alternative organizational assumptions for human behavior.
>
> (Simon 1996: 21)

Simulating human problem-solving through computer models thus enables scientists to gain knowledge about the functional dynamics of the modeled system. Simon (ibid.: 13) asserts that, "we try to understand the imitated system by testing the simulation in a variety of simulated, or imitated, environments." The problem here is to understand how the simulated system might respond to different kinds of interventions or provocations. Simulated disturbances create the disruptions that disclose systemic connections (cf. Chandler 2014a).

Of course, as the previous chapter demonstrated, simulation is a vital adaptive management technique for resilience proponents. Like Simon's studies of human cognition, simulations allow ecologists to test their understanding of system dynamics and visualize how a system might react to different management interventions (Holling 1978). Brian Walker and David Salt (2012) distinguish between conceptual models, which provide a basic understanding of how a system operates (see also Lee 1993), and analytical models, which "flesh out" conceptual models and enable scientists and practitioners to identify key thresholds. The point here is not to develop a totalizing representation that would enable predictive control, but rather to develop a functional understanding that allows managers to visualize how systems might react to different interventions and disturbances, learn from these reactions, adjust their interventions, and manage for sustainability (Holling 1978). Mirowski characterizes Simon's use of models in similar terms: the latter is "quite happy to continue to 'simulate human behavior over a significant range of tasks, but do not pretend to model the whole mind and its control structure'" (Mirowski 2002: 463, citing Simon 1991). Simulations thus do not produce objective, analytical knowledge but instead produce functional, synthetic knowledge: they enable interventions in complex environments, and learning from these interventions. And because they involve nothing more or less than information exchange, scientists can use them to develop functional understandings of any complex system, whether this is the human psyche or large-scale social and ecological systems.

Simon's sciences of the artificial thus offers an informatics-oriented vision of adaptive behavior in complex environments that drags psychology, economics, public administration, design studies and, indirectly, ecology into the cybernetic age. In following the reductionist path cyberneticians mapped out, Simon engages in the same practices of rational abstraction that reduce life to information. But I want to pause here and situate this practice of abstraction and functional re/combination within its wider historical and cultural context. Both Simon and his fellow cyborg scientists were products of a particular historical moment: the emergence and consolidation of a new aesthetic – or sensory – experience of the world as a systemic environment comprised of functional objects that could be combined and recombined with one another in a variety of ways. The next section situates Simon's work specifically, and cybernetics more broadly, within the aesthetic regime of modernist design consolidated through the Bauhaus movement.

MODERN DESIGN AND THE AESTHETICS OF CYBERNETICS

Few writers convey early-twentieth-century Europe's spirit of dislocation better than critic Walter Benjamin. In his enigmatic "Theses on the Philosophy of History," he famously suggests that, "the tradition of the oppressed teaches us that the 'state of emergency' in which we live is not the exception but the rule" (1968a: 257). Subsequently, scholars have departed from this observation to explore how the formal and informal emergency structures everyday life (e.g., Taussig 1991; Agamben 2005; see Chapter 7). However, for our purposes here, Benjamin's observation illustrates the foreboding sense of catastrophe that permeated early-twentieth-century European life. He lived and wrote in a time of cultural, political and economic upheaval, and his writing signals a pervasive sense of crisis and groundlessness: he described himself as, "like the one who keeps afloat on a ship wreck by climbing to the top of a mast that is already crumbling" (Benjamin 1931, in Arendt 1968: 19). The killing fields of World War I had transformed the Enlightenment's boundless promise of technology and rationality into boundless violence and revolutionary upheaval. Indeed, philosopher Peter Sloterdijk (2009) has detailed the meticulous scientific calculations military planners undertook to launch the first chemical warfare attacks in 1915. Killing on an industrial scale occurred through what he called the *explication* of the environment. By this term, Sloterdijk refers not only to the reduction of the sensible world to an objective reality that can be measured and known with quantitative precision – a mode of apprehension that dates back to at least Descartes (Elden 2006). More specifically, *explication* signals an apprehension of the environment as a series of *abstractions* that can be *functionally aligned* with one another *to produce systemic outputs*. In the case of chemical warfare, creating a "death environment" required understanding the movement of poison gas clouds, a flow that could be impacted by atmospheric conditions such as temperature, humidity, wind speed and direction. And it required knowledge of what concentrations of gas in the air could be merely damaging to human physiological functioning, and what concentrations would be lethal – which in turn required knowledge about how different gases interacted with respiratory, circulatory, and nervous systems in living, breathing organisms. Armed with this knowledge, military planners and researchers could create mathematical formulas representing functional relationships between abstract variables, which allowed them to determine the approximate concentration a gas cloud would be at a target destination if they released poison gas canisters at point x in certain wind, atmospheric, and temperature conditions. They could then determine how many canisters of gas they would need to create the intended lethal environment.

The advent of chemical warfare and the industrial production of killing is one of the more visceral expressions of the wider political, cultural, economic and social crisis that saturated the twentieth century's first decades. If the Enlightenment had hitched a conservative social order and its attendant hierarchies of meaning and value to a secularized form of salvation – the promise of unlimited growth and development, facilitated through technology and rationality – then Europe's crisis demonstrated how rationality and technology could be implicated in the most unimaginable forms of violence.[14] The catastrophic collapse of the prevailing order left a void that revolutionary movements sought to fill as they struggled with questions of how to produce meaning and value in the absence of transcendental guarantees. In the realm of politics, Bolshevik revolutionaries promised a new Soviet state founded on the immanent proletariat rather than the transcendent "nation." Their German counterparts took to the streets of Berlin and overthrew the Kaiser, even as their revolutionary aims eventually ran aground in the liberal Weimar constitution. In the realm of art, abstract artists such as Wassily Kandinsky experimented with forms of expression that were anchored in the artist's own subjective feelings rather than reference to external nature. Dadaists in Zurich led by Hugo Ball (1974) experimented with language and sound in the absence of a centering human subject to ground meaning. Their performances pushed meaning beyond its limit, mimicking and exaggerating the subjective effects of military collapse, political crisis, mechanization, commodification and massification (Foster 2015). And artists and writers such as Benjamin (1968b), Theodor Adorno, and Ernst Bloch faced the problem of how art and technology might merge together in new forms of sociable life.

This discussion might appear far removed from cybernetics, mid-century behavioral science, and ecological resilience theory, but this period of crisis also gave rise to the designerly mode of apprehending the world each shares in common. In the midst of this tumult, in 1919 the German architect Walter Gropius founded the Bauhaus art school in Weimar, Germany. The Bauhaus sought nothing less than to refashion the boundaries of art and technology. Gropius (1919) cast aside entrenched social hierarchies that "raise[d] an artificial barrier between craftsman [sic] and artist," while artificially separating different forms of art such as painting, sculpture, architecture and theater. Instead, he sought to produce a newfound unity of the arts – what he called the "complete building" – through a new mode of artistic instruction centered on the workshop (Koehler 2009; Rüedi Ray 2010). The Bauhaus' program for training craftsmen and women[15] consolidated ongoing modernist artistic movements that privileged function over form. The latter was associated with class hierarchy, and specifically the ornately decorated furniture, housewares and buildings of the elite. Plebian efforts to copy these ornate works flooded mass markets as cheaply manufactured kitsch (Foster 2002), sparking modernist critics such as Adolf Loos (1997) to decry ornamentation as a waste of labor, health, material and capital. The Bauhaus fed off these sentiments to

Figure 5.1 Olivetti Studio 42 typewriter, Alexander Schawinsky (ca. 1936). Schawinsky's typewriter expresses a modernist design aesthetic that privileges function over form. His interpretation reduced the artifact to essential abstract shapes – circles, lines, curves, rectangles – that together formed a functional system: an object that could write characters.

Image from Wikimedia Commons.

reconceptualize everyday products, such as furniture, clothing, architecture, or typewriters (see Figure 5.1), as nothing more than embodiments of the function they were meant to perform in daily life – thus stripping away all ornamentation to reveal objects in their pure functionality (Foster 2002; Groys 2008).

The Bauhaus thus reduced art to life, and in doing so, gave birth to modern design. By emphasizing the unifying function of craftwork, it brought art down from the heavens and folded it into the design of artifacts for everyday living. However, in the same move, it elevated everyday life (and importantly, the functional objects that comprise everyday life) to a work of art (Foster 2002; Schuldenfrei 2009). This is an artistic expression of the logic of explication, as above, which enframes the world as rational abstractions given to functional synthesis. For philosopher Jean Baudrillard (1981), the Bauhaus' recalibration of the relation between art and life marks a seismic shift in the way individuals apprehend one another and the wider world. In his reading, the Bauhaus "liberates functionality" and allows objects to form systems with other objects based on functional coherence: "the whole environment becomes a signifier, objectified as an element of signification. Functionalized and liberated from all traditional implications (religious, magical, symbolic), it becomes the object of a rational calculus of signification" (ibid.: 186–187). The Bauhaus consolidates an experience of world as "a rational conception of environmental totality" (ibid.), or a functional system of objects amenable to human intervention and manipulation. And this experience involves a "dissociation of every complex subject-object relationship into simple, analytic, rational elements that can be recombined in functional ensembles and which take on the status of the environment" (ibid.).

Baudrillard's argument here draws attention to the aesthetic regime that underpins design theory – and thus cybernetics and resilience. By *aesthetic*, I follow Jacques Rancière (2004: 8) who characterizes the aesthetic as "the system of *a priori* forms determining what presents itself to sensory experience. It is a delineation of spaces and times, of the visible and the invisible, of speech and noise, that simultaneously determine the place and the stakes of politics as a form of experience." The aesthetic thus orders sensory experience: it refers to partitions, or distributions, of sensible experience that assign people and things to "proper" spaces and times. For our purposes here, Rancière's (2009, 2013) genealogies of the modern aesthetic show that sensory experience is historically contingent. There is nothing inherently natural or pre-determined about the way we sense the world, orient ourselves, and judge the value of other people and things. Instead, sensory experience is an effect of power relations that give each person and thing their "proper" space and time (Grove and Adey 2015). Although Baudrillard does not use Rancière's terms, the "*dissociation* of every complex subject-object relationship into simple, analytic, rational elements…" referenced above indicates a certain aesthetic, a partitioning of sensible experience into functional abstractions.

The experience of environmental totality is thus an aesthetic experience. But importantly, as Baudrillard emphasizes, this experience is the precondition of both cybernetics and design as Herbert Simon defines it: to synthesize interior and exterior through the production of artifice requires *an experience of interior and exterior as objective entities amenable to functional synthesis*. The Bauhaus thus crystallizes a cultural-historical transformation of the modern aesthetic that had been emerging across a number of scientific, artistic and political fields in response to the early twentieth-century's pervasive sense of catastrophe.

We can illustrate this transformation by contrasting the modernist *design* aesthetic with Rancière's genealogy of the *modern* aesthetic. Rancière (2013) focuses on the period surrounding the European Enlightenment, when the modern aesthetic emerged as an inversion of Enlightenment rationality. If the modern subject, after Kant, was a self-consciously rational subject divided from an objective world, then the aesthetic subject was always a part of the world, immersed in sensuous experience (Eagleton 1990). In contrast to the objective truth of deductive scientific analysis, the aesthetic offered subjective judgment on beauty and taste. Rancière shows how the aesthetic orders sensory experience in a way that shores up rather than undermines the possibility of objective truth. It brackets subjective, embodied experience, and prescribes a certain technique – art – for engaging this domain. In contrast, Baudrillard's analysis highlights how design and cybernetics blur beauty and truth through functionality in a way that transforms the aesthetic:

> for every form and every object there is an objective, determinable *signified* – its function. This is what is called in linguistics the level of *denotation*. The Bauhaus claims to strictly isolate this nucleus, this level of denotation ... the thing denoted (functional) is beautiful, the connoted (parasitical) is ugly. Better yet: the thing denoted (objective) is true, the connoted is false (ideological). In effect, behind the concept of objectivity, the whole metaphysical and moral argument is at stake.
>
> (Baudrillard 1981: 196)

Design after Bauhaus thus pushes the modernist pursuit of objective truth to its limit, where it becomes indistinguishable from a modernist understanding of subjective beauty, likewise pushed to its limit. At this threshold lies a "functional calculus" that extends to "the whole field of everydayness" (ibid.: 193): Canguilhem's (1994) "non-numerical calculus" referenced above.

This functional calculus has at least two significant effects that are relevant for understanding design and resilience. First, it splits both subject and object into sign and signified at one and the same time. The subject no longer grounds objective truth, as in Descartes; rather, its own truth is always in question. The subject sits at the threshold between self-image and self-as-image: her own sense of self, or the qualities, characteristics and values she possesses; and the judgmental gaze of others that affirms or denies those qualities, characteristics and values, respectively. In the age of modern design, the subject presents herself as object to be judged by others, even as she participates in judging others: "individuals appear both as artists and self-produced works of art" (Groys 2008: 6). She is always caught up in the process of (self-)design, adapting and aligning internal self-image with external judgments. Truth becomes an emergent quality determined by particular alignments between interior and exterior.

Second, a functional calculus also transforms the possibilities for control. If modern science promised to tame an ontologically separate nature through predictive knowledge, the emergent truth of design can offer no such guarantees. Instead, design offers the potential for pragmatic synthesis. As a functional calculus envelops wider swaths of everyday life, material entities – both subjects and objects – become little more than signifiers, or signs, denoting signifieds, which after Bauhaus consist of abstract functions entities can perform. This is the cultural experience of environmental totality: we are confronted with a world made up of objects – people, non-human organisms, material objects, technologies, and so forth – that we apprehend as potential functions. As we saw above, this experience is the precondition for Simon's epistemology of simulation. But it is also the point where design becomes a real possibility: "this split, this fundamentally broken and dissociated relationship ... between man [sic] and his environment is the *raison d'être* and the site of design" (Baudrillard 1981: 193). Control becomes a matter of suturing this split through the pragmatic act of synthesis: combining functions in unique ways in order to realize specific goals. And as we saw above, synthesis is the location of artifice, the feedback-driven adaptation between interior and exterior, and thus the site of truth.

The design aesthetic thus transforms the relation between truth and control. Design does not reduce the world to objective entities that can be known with precision and thus folded into calculative and predictive rationality. It does not mobilize a will to truth. Instead, design renders entities as functional abstractions that can be pragmatically synthesized with one another in order to solve specific and pressing problems. This striving to strategically align interior and exterior expresses a *will to design*: the effort to control human and non-human entities through adaptive synthesis rather than predictive analysis.

By now, we have traveled some distance from ecological research on resilience. And yet, it remains close by: despite the various transformations, translations, and discontinuities between design, cybernetics, Simonian behavioral science and ecological resilience, they all mobilize, in their own way, a will to design that orients thought and practice around the imperative to reduce the world to rational abstractions that can be functionally synthesized with one another. Each field of study circulates within the same aesthetic terrain that military planners, artists, scientists, writers and designers first surveyed in the early twentieth century. The next section returns to resilience theory to explore how this will to design inflects the production of ecological truth.

RESILIENCE AS DESIGN

If modern design promised nothing short of a revolution in the relation between art and life, cybernetics similarly celebrated itself as heralding a new era in scientific research. One cybernetician tellingly expressed this epochal shift in aesthetic terminology: "now, after the age of materials and stuff, after the age of energy, we have begun to live the age of form" (Auger, quoted in Bowker 1993: 111). Thus, as Geoff Bowker (ibid.) suggests, "cybernetics as the science of form [rather than content] could, then, replace materialism as the philosophical avatar of the political economy." And this new science promised a new handle on seemingly runaway problems of complexity. If life can now be grasped in formal terms as functional abstraction, (certain) humans can create new life through genetic engineering and destroy the world through atomic warfare. Cybernetics promised to navigate these turbulent waters and achieve peace and prosperity through feedback-mediated control and regulation (DeLanda 1991). More than a set of techniques, it involved a mindset or a way of apprehending the world as environmental totality, and grasping the new possibilities for control and regulation this created. Another cybernetician suggested in language worth quoting at length that,

> control theory, like many other broad theories, is more a state of mind than any specific amalgam of mathematical, scientific or

> technological method... As mathematicians, physiologists and engineers explore the subtle difficulties of dealing with large-scale systems, it seems less and less likely that any single "cybernetic" theory will serve all purposes. Nevertheless, for those who want to understand both modern science and modern society, there is no better place to start than control theory"
>
> (Bellman 1964, quoted in Bowker 1993: 116–117)

While there is no panacea (Ostrom *et al.* 2007) for managing large, complex systems, cybernetics provides an entry point, what Bowker (1993) calls a distributed passage point, for scholars from any number of fields seeking to understand systemic dynamics.

Resilience scholars discuss their craft in similarly epochal terms. Fridolin Brand and Kurt Jax (2007: unpag.) note that, "resilience is increasingly interpreted in a broader meaning across disciplines as a way of thinking, a perspective or even paradigm for analyzing social-ecological systems." Carl Folke (2006: 260) suggests that,

> resilience is an approach, a way of thinking, that presents a perspective for guiding and organizing thought and it is in this broader sense that it provides a valuable context for the analysis of social–ecological systems, an area of explorative research under rapid development with policy implications for sustainable development.

Brian Walker and David Salt (2012: 1) explicitly couch this epochal shift in aesthetic terms: they suggest that a "resilience frame of mind" is a prerequisite for managing complex social and ecological systems. This frame of mind includes, *inter alia*, heightened awareness to systemic interconnections and relations, increased sensitivity towards the existence of thresholds that can rapidly tip a system from one stability regime to another, willingness to accept change and live with risk and vulnerability, and an interest in treating change, crisis and insecurities as opportunities to learn and adapt (Grove and Adey 2015).

Like cybernetics and Simon's sciences of the artificial, a resilience frame of mind apprehends the environment as rational abstractions reducible to pure functionality. In order to experience nature as a complex ecosystem – to think in terms of interconnections, emergence, feedback and thresholds – the resilience scholar first needs to be in a "frame of mind" that perceives objects as de-contextualized signs that denote functionality. This is the same sensorial experience that conditions cybernetics and Simon's sciences of the artificial. As Baudrillard (1981: 190) puts it:

> objects, forms and materials that until then [the Bauhaus] spoke their own group dialect, which only emerged from a dialectical practice or an original style, now begin to be thought of and written out in the same tongue, the rational Esperanto of design. Once functionally liberated, they begin to make signs, in both senses of the phrase ... that is, they simultaneously become signs and communicate among themselves. Their unity is no longer that of a style or practice, it is that of a system.
> (Baudrillard 1981: 190)

A modern design aesthetic thus makes the development of systems thinking possible in any number of fields, for it reduces the world – "what exists" – to abstract, de-contextualized bits of information. Resilience, cybernetics and Simon's science of the artificial all think and speak the universally applicable "rational Esperanto" of design.

This means that the complex ecosystems resilience addresses are neither the produced "second Nature" (Smith 1984) that geographers and political ecologists have explored since the 1980s nor the ontologically pure "first Nature" of modernist ontology (Latour 1993). These visions of nature result from dialectical analysis and a metaphysical concern with an original style, respectively. Instead, resilience thinking perceives nature as the post-Natural, abstracted, functionalized and systematized "environment" of cybernetics and design. Baudrillard (1981: 202) explains that:

> If one speaks of environment, it is because it has already ceased to exist. To speak of ecology is to attest to the death and total abstraction of nature ... the great signified, the great referent Nature, is dead, replaced by environment, which simultaneously designates and designs its death and the restoration of nature as simulation model (its "reconstitution," as one says of orange juice that has been dehydrated).

Baudrillard's argument about the "death of nature" is a philosophical claim. It refers to the reduction of being – "what exists" – to signs designating entities' functionalities. This abstracts existents from the contexts from which they emerged and the web of affective relations they are enmeshed in. Thus reduced, "what exists" can be endlessly folded into simulation models that attempt to strategically control and manipulate "environments" to produce certain outcomes.[16] In political-philosophical terms, this reduction of nature to environment and its reconstitution in ecological simulation models negates potentiality. That is, potentiality – the affective potential to become radically otherwise to oneself that saturates everyday encounters (Grosz 2009; Tsing 2015) – becomes fully actualized as the determinate possibility to be functionally re/combined with other entities. To experience "environment" as such is to experience world as open to endless possibilities of cybernetic control and regulation – and paradoxically, devoid of the indeterminate potential to become radically otherwise (Evans and Reid 2014; Grove 2017).

Simon's understanding of environment exemplifies this death of nature. In his key essay on the choice environment that bounds rational decision-making,[17] Simon (1956) defines "environment" in terms Baudrillard might immediately recognize:

> the term *environment* is ambiguous. We are not interested in describing some physically objective world in its totality, but only those aspects of the totality that have relevance as the "life space" of the organism considered. Hence, what we call the 'environment' will depend upon the needs, drives, or "goals" of the organism, and upon its perceptual apparatus.
>
> (Simon 1956: 130)

This is environment as a cybernetic space, a contingently relational space that draws together elements on the basis of their functional utility. Its contingency reflects the fleeting nature of these elements' functions: their possible utility is determined by an organism's specific "needs, drives or 'goals,'" as above. The roots of behavioral engineering lie in this vision of environmental totality (e.g., March 1977). Indeed, while the interior psychological processes that make up individual choice may lie outside the reach of totalizing scientific knowledge and direct control, the external environment can be reduced to a series of functional abstractions that shape how individuals make their choices (Jones *et al.* 2013). Design enters here as a way of thinking about how to align internal psychological processes with external environmental cues. The science of the artificial, in this case, becomes a cybernetic science of social control: artifacts – norms, institutions, procedures, rules, programs, incentives, and so forth – can shape the quality of feedback between internal and external environments, and thus direct, but not determine, individual choices.

Simon's cybernetically inflected science of the artificial thus produces new means of regulating individual and collective life *through* the liberal freedom of individual choice. Left to their own devices, individuals may pursue their goals and interests through a variety of strategies and actions that, to observers, appear as complex behavior. But the provision of well-designed artifacts can guide their strategies and actions in ways that produce specific outcomes. David Chandler (2013, 2016b) has picked up this genealogical thread in his analysis of behavioral governance in the United Kingdom. He demonstrates how Simon's sense of how the environment shaped individual decision-making processes influenced Friedrich Hayek's psychological research, Anthony Giddens' "Third Way" approach to politics, and the current interest in so-called "paternalistic" governance (see also Jones *et al.* 2013). These all envision governance of complex social and ecological problems as a process of engineering the "choice environments" that confront individual citizen-consumers. Rather than imposing controversial mandates – for example, measures to increase environmental sustainability – the manipulation of choice environments guides individuals to make new behavioral decisions seemingly on their own.

Resilience thinking extends this cybernetic vision of "environmental" engineering to ecological processes. Discussing the value of Simon's theory of hierarchy, Holling (1978: 28) suggests that,

> Simon has shown that such [hierarchical] structures have remarkable survival properties. First, removal of one sub-assembly does not necessarily destroy the whole. Because of the minimal connection between subassemblies, the others can persist, often long enough for self-recovery. Second, for the same reason, these structures rapidly adapt to change. As long as the same connections are maintained to other subassemblies, major changes and substitutions can take place within the sub-assembly. Species can substitute for species as long as the same function or role is performed.

In ecologists' hands, species become reduced to abstractions that can functionally substitute for one another. What matters for resilience thinking is *not* the precise identity of individual elements within the system. This is the concern of an analytical natural science, or reasoning in terms of dialectics or style to borrow Baudrillard's terminology from above. Instead, resilience thinking is concerned with the functional roles of species – and multiple species can serve the same functional role. To be sure, ecologists often mobilize this sensibility to argue for greater diversity: a diverse ecosystem, at any scale, will be resilient because other species can fill in and perform the same functions, and thus ensure systemic meta-stability, even as the system changes form and identity (Walker and Salt 2012). But diversity has these effects because species are nothing more than functional abstractions that can potentially substitute for one another.

An aesthetic of environmental totality thus lies at the heart of resilience proponents' transformations of scientific and environmental management practices. Just as Simon's sciences of the artificial reconfigure social science around a cybernetic sensibility and its underlying will to design, the "very different view of the world" Holling (1973: 15) offers recalibrates ecological science and environmental management around an aesthetics of environmental totality. The result is a new form of scientific and management practice that extends Simon's recuperation of rationality to the study of human–environment relations. The concluding section explores this science of ecological artifice.

THE SCIENCE OF ECOLOGICAL ARTIFICE

In an edited volume that helped lay the groundwork for resilience scholars' engagement with institutional economics, Holling characterizes ecological resilience as a distinct form of science: not a "science of parts" that provides detailed analytical knowledge of each element within the system, but a "science of the integration of parts" (Holling 1995: 13). As the name suggests, this science explores the functional relations that determine system performance. It is a science of thresholds – the interface between, on the one hand, an "interior" realm of environmental science and management, and on the other, an "exterior" realm of complex social and ecological systems. And it transforms conventional scientific and management techniques in order to synthesize interior and exterior. As Holling explains, a science of the integration of parts "uses the results and technologies of the [science of parts] but identifies gaps, develops alternative hypotheses and multivariate models, and evaluates the integrated consequences of each alternative by using information from planned and unplanned interventions in the whole system that occur or are implemented in nature" (Holling 1995: 13). In essence, Holling is describing a science of ecological artifice that re-purposes conventional scientific techniques of modeling around the cybernetic imperative to tease out functional relations and manage feedback loops in order to improve system performance – in this case, by enhancing resilience and building adaptive capacities.

If cyberneticians could frame their theory as a unique angle on the "subtle difficulties of dealing with large-scale systems" of all types (Bellman 1964, quoted in Bowker 1993: 116–117), resilience scholars likewise position their new form of scientific practice as a response to the challenges of complexity. Holling (1995: 13) argues in Simonian language that,

> the premise of this second stream [a science of the integration of parts] is that knowledge of the system we deal with is always incomplete. Surprise is inevitable. Not only is the science incomplete, but the system itself is a moving target, evolving because of the impact of management and the progressive expansion of the scale of human influences on the planet.

Holling offers here a vision of second-order cybernetics or general complexity: the systems ecologists attempt to manage are not closed-loop systems of servomechanism theory, but are instead open systems that interact with their surrounding environment. There is no position from which scientists and environmental managers can attain totalizing, objective views of the system. Knowledge of the system is thus always limited (in Simonian terms, bounded) and emerges through the stepwise, adaptive process of interacting with the system: through designing interventions (for Simon, artifacts), monitoring how those interventions affect system performance, and adjusting interventions based on the new information.

Set in a context of cybernetics and its aesthetic of environmental totality, resilience becomes the means for synthesizing diverse forms of bounded knowledge around specific problems of social and ecological complexity. This is the aesthetic register of Brand and Jax's (2007) description of resilience as a boundary object. As they note, boundary objects "facilitat[e] communication across disciplinary borders by creating shared vocabulary although the understanding of the parties would differ regarding the precise meaning of the term in question" (ibid.: unpag.). Resilience is an effective boundary object because, as we have seen, it ultimately does not refer to specific natures, environments, or social groups, but rather to de-contextualized rational abstractions. *Resilience is an empty signifier that signals nothing other than potential for functional synthesis* – and this cybernetic potential is the basis for what Brand and Jax (ibid.) call a "transdisciplinary approach to analyze social and ecological systems."[18]

The techniques of resilience the previous chapter detailed facilitate this transdisciplinary synthesis. Adaptive management and adaptive governance techniques offer the promise of de-centering scientific privilege, even if this does not always play out in practice (see Nadasdy 2007; Rist *et al.* 2013). The design of new institutional arrangements – or new artifacts – through the adaptive management process can produce a "new role" for scientists as "one of several actors in the learning and knowledge-generation process" (Folke *et al.* 2005: 445). A science of the integration of parts thus works against scientists' "natural" tendency to engage in narrow, analytical reasoning. Indeed, as Holling cautions, scientific analysis often has limited utility for managers and other resource users:

> Another design factor is the potential danger of confusing intellectual curiosity with practical information needs. It may please the analyst to know that he [sic] has correctly predicted how a population will respond to excess harvesting, and the experience accumulated may be useful in future cases. But the direct beneficiaries of the resource may quite rightly have little sympathy with these results; they will be concerned with the narrower empirical problem of how system outputs of immediate value to them are related to policy inputs, and with the risks of any experimental probing relative to the improvements that probing may uncover.
>
> (Holling 1978: 181)

Instead, "research necessary for adaptive assessment and design must be focused through policy concerns" (ibid.: 4). This new focus allows adaptive management to synthesize different forms of knowledge in order to produce functional knowledge of systemic performance – the kind of functional knowledge that will be useful to managers and other practitioners seeking to achieve sustainability. In resilience proponents' hands, environmental management becomes a Simonian process of designing adaptive management procedures that enable scientists to set aside their narrow disciplinary concerns and collaborate with other scientists and non-scientists to create satisficing solutions.

Ecological resilience theory thus institutes a cybernetic will to design at the heart of scientific practice. It transforms the object of scientific inquiry into rational abstractions that can be functionally synthesized with one another. And it transforms scientific analysis around the imperative to synthesize different, bounded forms of knowledge in order to develop pragmatic solutions to problems of complexity. Holling's science of the integration of parts is a science of ecological artifice that explores how particular alignments between knowledge and nature, policy and output, generate certain kinds of systemic capacities. It explores the feedback-driven links between system components, and the effects of these links. In Simonian terms, it sits at the threshold between interior and exterior. It is a cybernetic process of ecological design, the process of adapting the interior of scientific knowledge and policy with the exterior of complex social and ecological dynamics.

At stake in a science of ecological artifice is the possibility of rationality in environmental management. As we saw above, Simon's theory of bounded rationality transforms rationality into adaptation: rational behavior is that which realizes goals within a complex world through a stepwise search for satisficing solutions. In Holling's science of the integration of parts, resilience substitutes for adaptation: a resilient system is structured through feedback mechanisms that allow its interior to adaptively align with its exterior. The environmental management process produces artifacts – policies, interventions, strategies and so forth – that work on these feedback mechanisms to ensure interior and exterior remain aligned. Resilience thus names the form rational behavior assumes in conditions of social and ecological complexity. To be resilient is to be rational in a world of bounded rationality. The resilient system and the resilient individual are both capable of achieving goals through the process of adapting to complexity. These goals might be conservative, such as preserving functional performance, or they might be more progressive, such as transformative change to achieve sustainability. But they all exhibit the adaptive behavior that Simon designates as the new, cybernetic rationality.

Ecological resilience thus does not *break* with conventional forms of environmental management as much as it attempts to *recalibrate* environmental management for a world of complexity. It retains the underlying belief that management should be based on rationality. But this rationality is now a Simonian rationality of inductive, adaptive, goal-seeking behavior rather than the deductive application of expert knowledge. And as this chapter has shown, this vision of rationality emerges out of an aesthetic experience of environmental totality, which for resilience proponents amounts to sensing the world through a "resilience frame of mind" (Walker and Salt 2012: 1). The truth of systemic components – social or ecological – is their functionality; generating this truth requires cross-boundary work that synthesizes different forms of knowledge – both scientific and lay – on complex systems; and folding this truth into management requires techniques of adaptive management and adaptive governance that can create artifacts – procedures, policies and institutions – that synthesize these diverse forms of knowledge.

This chapter has thus shown how Herbert Simon's influence on formative research into ecological resilience leads us to the aesthetic dimensions of resilience. Resilience thinking introduces an aesthetics of environmental totality to the study of human–environment relations: it reduces these relations to an interface between interior and exterior, and transforms scientific study into a pragmatic and collaborative exercise of manipulating feedback loops that structure this interface. Rather than a will to truth, resilience approaches mobilize an ecological will to design that attempts to create social and ecological systems with certain desirable characteristics despite, or even *through*, necessarily limited knowledge.

The genealogy of resilience offered here thus opens up qualitatively distinct ethical and political questions. A will to design *intensifies* a liberal will to truth's efforts to fold the totality of social and ecological phenomena into a calculative governmental rationality. As we saw above, a will to design concerns itself with what something can do rather than what it is. In other words, a will to design operates on embodied, dynamic and relational potentiality rather than static actuality. In the process, the terrain of control shifts to intimate, affective relations that structure potentiality: control no longer operates through predictive certainty but rather inflects the choice environment that influences, but does not determine, how individuals act. The next chapter examines how resilience approaches attempt to govern affective potential.

FURTHER READING

Although Herbert Simon deals with somewhat arcane theoretical and methodological material, his writing is lively, accessible and often enjoyable to read. His autobiography *Models of My Life* provides a firsthand account of his various exploits across the disciplines, and *The Sciences of the Artificial* contains the most concise summary of his work available. Importantly, the third edition (1996) contains many updates that incorporate developments in design studies that critiqued earlier field-defining editions of the text. The links between his work and resilience-based adaptive management approaches are particularly evident here. Simon's work drew on debates in a wide variety of fields across philosophy and the social sciences, which makes contextualizing his *oeuvre* a daunting task, but Hunter Crowther-Heyck's *Herbert Simon: The Bounds of Reason in Modern America* provides a remarkable account not only of Simon's work but also the development of twentieth-century American social science in general. Philip Mirowski's monumental *Machine Dreams: Economics Becomes a Cyborg Science* drills down into the transformations in Western economic thought during the early and mid-twentieth century, which also helps situate Simon's work in relation to the discipline's broader cybernetic turn.

NOTES

1 Here, it bears noting that one of the strengths and weaknesses of much critical scholarship on resilience is the easy elision between resilience and neoliberal rule. On the one hand, this work has helped to denaturalize many of resilience proponents' more inflammatory claims, such as the cybernetically inspired aspiration to provide a universally valid account of social and ecological change. But on the other hand, much of this critical scholarship passes over what needs to be explained: how it is that disconnected and diametrically opposed critiques of centralized governance came to resonate with each other. As Chandler (2014a) shows through his discussion of resilience as a "resonance machine," any alignment between resilience and neoliberalism is a contingent accomplishment rather than an ideological necessity. The genealogy of resilience is considerably more complex than many first-cut critiques allow, and several scholars have begun to offer important correctives and interventions that trace out alternative genealogical narratives (Chandler 2014a; Schmidt 2015; Cavanagh 2016; McQuillan 2016).

2 There are many similarities here between Baudrillard's understanding of environmental totality, Peter Sloterdijk's (2009) arguments on environmental "explication" and Martin Heidegger's (1996) sense of "environmentality." In their own way, each signals a phenomenological experience of the world as an objective reality given to human experience as a functional system. For example, to explicate the environment, in Sloterdijk's sense, is to render surrounding phenomena, such as air, wind, rain, and so forth as objective conditions that lend themselves to human intervention and manipulation, such as air pressure, humidity, temperature, and wind speed. However, Baudrillard is unique in locating the emergence of design as a pivotal moment in the cultural shift all three thinkers signal in one way or another. At the same time, Baudrillard's thought is problematically distinguished by a totalizing vision of power that fully envelops life – a vision of "the political economy of the sign" that slowly extends into the fabric of everyday life and transforms subjectivity in accordance with the dictates of capitalist political economy. My intention in this chapter is thus to bracket these totalizing aspects of his thought and focus instead on his characterization of environmental totality and the links

he draws with design. As I will suggest later in this chapter, I find Jacques Rancière's (2004, 2009) Foucauldian-tinged understanding of aesthetic regime a more useful way to think through the transformations Baudrillard signals – particularly because Rancière allows for elements of contestation and indeterminacy within specific configurations of the modern aesthetic regime.

3 One of Simon's fiercest critics was Hubert Dreyfus, who argued that Simon's repeated assertion that computers could think once they could be programmed to simulate human problem-solving strategies hinged on a narrow understanding of human intelligence (Dreyfus 1965). In particular, Dreyfus argued that Simon's work on artificial intelligence – particularly his collaborations with Alan Newell – presupposed a certain kind of human mental activity restricted to very precise and abstract situations, such as solving mathematical problems or playing chess – two of Simon and Newell's favorite activities to simulate. While they succeeded in more or less adequately simulating this narrow range of intellectual activity, Dreyfus argued that much of what makes up human intelligence is, first, poorly understood in analytical or synthetic terms, and second, consists of practical, day-to-day know-how. See Sent 1997; Mirowski 2002; Crowther-Heyck 2005 for extended discussion of these debates.

4 First-order cybernetics is roughly akin to what we have been discussing in David Chandler's (2014a) terminology as "simple complexity." It envisions a closed system formed through complex interconnections or feedback loops between elements. A classic example is the servomechanism, a control device that automatically corrects for the error between a machine's input and an output. Servomechanisms, such as thermostats or radar-guided missiles, provided the practical basis for cybernetic understandings of adaptive behavior and change in social systems (more on this later in the chapter).

5 While Simon's theory of bounded rationality was formally developed in two publications focusing on the internal (Simon 1955) and external (Simon 1956) limits on rationality, his influential and Nobel Prize-winning *Administrative Behavior* (Simon 1997) developed the theory's basic ingredients, even if it did not explicitly spell them out as such.

6 We could just as easily argue that the fire department is a lower level within hierarchically structured municipal government, which is itself tasked with the overarching goal of ensuring local economic development, quality of life, and so forth.

7 On this point, the influence of John Dewey and William James' pragmatist philosophy, Talcott Parsons' functionalist sociology, and John Commons' institutional economics is particularly clear. From Dewey and James, Simon (1997) takes the point that individual psychology is shaped at the interface between mind and world, where the world continually presents the individual with situations she must work her way through. Habit is essential for engaging with a complex world, because it allows the individual to differentiate between stimuli that demand her immediate attention and stimuli that she knows, through prior experience, are not pressing. From Parsons and Commons, Simon takes the point that institutions shape habits and expectations (Heyck 2008a). While some scholars note in passing the influence of

functionalist sociology on resilience theory (Olsson *et al.* 2014; Watts 2015), considerable work is needed to unpack these genealogical links and situate them within the context of inter-war sociology, ecology, and political economy (cf. Crowther-Heyck 2005; Heyck 2008a, 2008b).

8 As we will see in Chapter 7, Simon's thought here opens the door to questions of institutional design – in other words, what kinds of institutions will functionally align higher and lower levels of decision-making? And there are, of course, a number of parallels here between Simon's concern with functional integration, institutional economics, and resilience scholar's problem of institutional fit.

9 Simon does not use the term resilience, but he is clearly describing an adaptive quality very much like ecological resilience: the ability for a system of production to absorb shocks, reconfigure its activities, and thus adapt to disruption.

10 This is, we should note, a distinctly *liberal* and *humanist* understanding of institutions. Indeed, many institutional theorists assert that humans develop institutions in order to create a sense of certainty in an uncertain and indeterminate world (e.g., North 1994, 2005; Ostrom 1997). Uncertainty exists, in their formulation, because humans are uniquely free to pursue their interests and desires, improve their situation, and thus *become* different than they are. This freedom renders the future indeterminate, for there is no guarantee that pursuing these interests and desires will lead to the intended outcome. Becoming different (post-structural theorists, after Heidegger, Foucault, Deleuze, Butler, and others, might call this becoming-otherwise) thus carries a certain amount of insecurity, for one's interests in becoming different might not lead to intended results, or they may impinge on another's interests. Institutions emerge to manage this distinctly liberal problem of the relation between freedom and security (on the latter, see Foucault 2008): formal and informal laws, rules, regulations, and norms provide expectations for how others will behave, and thus introduce a degree of regularity to an otherwise indeterminate world.

11 The genealogy of cybernetics research is far more complex than this brief discussion indicates. Manuel DeLanda (1991), Philip Mirowski (2002) and Michael Dillon and Julian Reid (2009), for example, have all emphasized how distinct approaches to cybernetics are differentiated in terms of the value they attach to information and noise, the impact of unexpected information, and the role of the receiving audience in shaping the content and form of information, to name a few. However, given our interest in the cybernetic roots of Herbert Simon's work, which develop out of critical engagements with Wiener's slant on servomechanism theory (Mirowski 2002; Heyck 2008a), these wider developments remain on the sidelines.

12 Heyck (2008a) suggests that Simon's vision of adaptive behavior shares more in common with biologist W. Ross Ashby's account of organisms maintaining internal equilibria within a changing environment than Wiener's servomechanism theory.

13 Simon had access to the most advanced computing machines in the world through his consultancy work with the RAND corporation during the 1950s and 1960s. As a number of historians of science have noted, Simon's exposure to computers enabled him to formalize his theory of bounded rationality

and reconceptualize human decision-making as a cybernetic process of problem-solving (Sent 2000; Mirowski 2002; Heyck 2008a, 2008b).

14 Of course, this distinction between rationality and violence only held sway, if at all, in the colonial metropole. "Enlightened" life in the metropole hinged on arbitrary violence exercised on black and brown bodies in both colony and metropole (Stoler 1994; Hussain 2003).

15 Given the Bauhaus school's desire for inclusive enrollment, a large number of its students were women. However, given prevailing gender norms at the time, women tended to be steered towards gender-specific activities, such as weaving and small product production, rather than more physically demanding activities such as furniture production.

16 Baudrillard is signaling here a *tendency* towards limitless control, not an absolute condition. Giorgio Agamben's (2007, 2009, 2014) engagements with the concept of *dispositif* (typically translated as "apparatus") advance a similar argument, even as he problematically (Derrida 2009) locates the origins of *dispositif* in ancient Christian doctrine. For Agamben, the reduction of being to the rational abstraction of its "use" creates the precondition for total catastrophe: life itself becomes devoid of any inherent value or meaning, and can thus be put towards any use necessary – even if, as in Auschwitz, this use involves mass genocidal extermination in the name of racial purification. But it bears stressing here that there is no master, sovereign individual in the driver's seat: this tendency towards abstract use makes it possible for life to be put towards *any* purpose, even if the history of recent modernity has been one of violent mass extermination.

17 Simon's 1956 article is a companion to his 1955 article on bounded rationality – the former had to be cut and submitted to a different journal due to word length (Mirowski 2002).

18 It bears noting that Brand and Jax mobilize a somewhat narrow vision of boundary object. In their view, a boundary object exerts a unifying, centripetal force: it bridges otherwise insurmountable differences around a predetermined center. For Brand and Jax, this "center" is the pragmatic analysis of complex social and ecological systems. As a boundary object, resilience allows diverse forms of knowledge, both lay and scientific, to be synthesized in order to address specific problems of complexity. But other understandings of boundary object are possible. Indeed, in their influential paper that defined the term, Star and Griesemer (1989: 389–390) contrast the dispersed and indeterminate work of boundary object with the analytical "funneling" performed by the dominant sense of an "obligatory passage point" in much Actor-Network Theory. Anna Tsing follows this usage of the term in her important discussion of "freedom" among matsutake mushroom picker communities in Oregon. For Tsing (2015: 94), freedom as a boundary object is a "shared concern that yet takes on many meanings and leads in varied directions." It is this indeterminate sense of "varied directions" that Brand and Jax write out, for their use of boundary object offers only one direction: towards the analysis and sustainable management of complex systems. The distinction between determinacy and indeterminacy in both design theory and resilience thinking is an important point we will explore in Chapter 7.

6

RESILIENCE AS CONTROL

INTRODUCTION

Ecologists Brian Walker and David Salt (2012: 23) suggest in telling language that "resilience thinking is a problem-framing approach to your system that seeks to help you decide what's important for the sustainability of the things you value, that you should be focusing on." As the previous chapter detailed, this is Simonian rationality cast in terms of environmental management: for Walker and Salt, resilience is a higher-level heuristic device that simplifies environmental management in a complex world. However, in framing the problem of sustainability in some ways rather than others, Walker and Salt inadvertently demonstrate how resilience becomes a vector for power relations: any framing is partial, and something will always be excluded outside the frame. For example, Michael Dillon and Julian Reid (2009) show how the development of cybernetics gradually erased the context and subject of information exchange. Cybernetics first reduced language to communication (copying the logical positivists of the early twentieth century), and then reduced communication to information. The result was information entirely abstracted from context (Bowker 1993). In this light, the question becomes what kinds of exclusions resilience thinking produces as, in Walker and Salt's (2012: 23) language from above, it "helps you decide what's important for the sustainability of the things you value, that you should be focusing on." As we will see, resilience thinking's topological reconfiguration of rationality implicitly structures how we might understand problems of complexity: it focuses on aspects of complexity and sustainability that lend themselves to management through techniques such as adaptive management, adaptive governance, simulation, and participatory engagements. This means that resilience thinking is constitutively capable of thinking certain aspects of complexity – *and not others*. There is always a remainder, an outside, that resilience approaches cannot capture. As Nigel Clark (2010) notes in relation to climate change research, managerial framings of complexity are often incapable of engaging with the indeterminate potential of becoming radically otherwise.

However, with limited exception, resilience thinking has often struggled to adequately address such questions of power and politics. As Chapter 2 detailed, critics have long argued that resilience does little more than reinforce the political ecological status quo (see especially Nadasdy 2007; Davidson 2010; Voß and Bornemann 2011). Resilience-based efforts to empower marginalized and vulnerable populations often do little more than shore up existing political ecological inequalities that produce marginalization and vulnerability in the first place (Evans and Reid 2014; Grove 2014a). For their part, resilience scholars counter that techniques such as co-management, adaptive governance and transformation allow scholars to address "power sharing, decentralization and devolution of management rights, power asymmetries, and injustices that arise from structural inequalities of power" (Olsson *et al.* 2014, unpag.). Thus, for resilience proponents, resilience techniques *can* account for uneven power relations and their effects, even if they do not always foreground these considerations in their analysis.

In this chapter, I explore how this problematic relation between resilience theory and power results from the way resilience approaches subtly transform critique into a mechanism of cybernetic control. Traditionally, the task of critique has been to demonstrate the contingency of the present, to show how what is taken for granted could be other than it is, and thus detail how power relations stabilize and naturalize this contingent order of things. In contrast, for designerly thinking, all truth claims in a complex world are necessarily bounded and partial. The critical movement no longer lies in demonstrating contingency, but rather in the creative act of re-purposing what already exists. To be critical in a modernist design aesthetic is to create new meanings, functions and identities by playfully and experimentally recombining elements with one another.

This designerly slant on critique has important if often under-acknowledged implications for the politics of resilience research. Specifically, it enables resilience scholars to visualize key components of critique – especially conflict and subjugated knowledge – as forms of feedback that indicate institutional misalignment, and thus opportunities to (re)design institutions that mediate between complex social and ecological systems. The designerly roots of resilience thus create an ambiguous relation to critique and the potential to become otherwise. On one hand, resilience initiatives mobilize a designerly style of critique that attempts to generate new meanings, functions and identities out of nothing more than the existing arrangement of things. But on the other hand, this creative repurposing relies on transforming key categories of conventional critical thought into forms of cybernetic feedback. Rather than generating transgression, conflict and subjugated knowledge become indicators of design opportunities for resilience scholars to transform maladaptive socio-ecological systems into more sustainable and adaptive ones.

This chapter thus explores how resilience theory's designerly style of critique facilitates the capture and appropriation of social and ecological difference. Rather than destabilizing totalizing knowledge claims, the production and cybernetic manipulation of difference becomes the means of generating systemic stability in a complex and indeterminate world. To explore how resilience approaches attempt to capture the force of transgression, the next section charts the key components of designerly critique. I then detail the form of environmental power resilience approaches mobilize. This is a designerly power that can repurpose social and environmental phenomena – that is, create new meaning, function and value through playful and experimental recontextualization. The third and fourth sections examine how resilience theory repurposes conflict and subjugated knowledge, respectively, as mechanisms of cybernetic feedback that indicate institutional misfit, and thus design opportunities to transform complex social and ecological systems.

THE CRITICAL SENSIBILITY OF RESILIENCE

As Chapter 4 detailed, ecological resilience theory gained traction through sustained critiques of prevailing command-and-control approaches to environmental management. In general, this critical sensibility is a defining feature of European modernity. Michel Foucault (2007b) drives this point home in his reading of Kant's short essay, *Was ist Aufklärung?* (*What is Enlightenment?*). For Foucault, Kant expresses a certain "enthusiasm for revolution" attuned to the *inessential* in the present and the possibility that things not remain the way they are. This enthusiasm for revolution, rather than the event of revolution itself, is for Kant the marker of Enlightenment (Gordon 1986). It expresses a critical sensibility that questions the particular alignment of truth and power that sustains a given order. While Kant formulated his sense of critique against absolutist rule, and problematically sought to ground authority in a universal rationality,[1] in Foucault's reading, he more fundamentally opened the question of how we are to be governed. Kant's critical questioning of the present showed that configurations of truth and power are always contingent and potentially malleable. Any claim to truth is thus partial, and thus what we take for granted could always be other than it is.

This critical attitude has informed radical projects of all stripes ever since (see Box 6.1). For example, Karl Marx (1990) famously takes us behind the seemingly natural appearance of commodity exchange into the hidden world of commodity production where class exploitation becomes visible. Likewise, Foucault examines the constitution of modern subjectivity through critically analyzing regimes of truth in psychology, medicine, penal practice, anthropology, and sexuality (e.g., Foucault 1989, 1990, 1994). And Donna Haraway (1988) shows how the modern scientist's God's-eye "view from nowhere" is always situated – embodied, gendered, racialized, and sexualized – and thus inescapably partial. The list goes on. But despite various epistemological, ontological and ethico-political differences, these critical approaches all involve a common movement against totalization: each takes on commonsensical, accepted truths, and in doing so, creates new possibilities for transgressing a given social, economic and political order.

Box 6.1 CRITIQUE AND THE STUDY OF HUMAN–ENVIRONMENT RELATIONS

Since the mid-1970s, critical sensibilities have increasingly dominated geographic thought on human–environment relations (Castree 2004). Geographers' early engagements with critical social theory during the late 1960s and early 1970s laid the groundwork for this development (Gregory 1978; Cox 2014). David Harvey's (1973) seminal *Social Justice and the City* notably introduced geographers to Marxist analysis. Harvey's historical-geographical materialism showed how class relations shaped socio-spatial relations, and in particular the uneven geographical development of capitalism (Harvey 1985; Smith 1984). This critical sensibility soon infected hazards and vulnerability studies, which had until then been dominated by Gilbert White's (1945) pragmatic behavioralism (see also Kates 1962; Burton *et al.* 1978). Where this latter work emphasized risk perception and decision-making (and was indeed influenced by Herbert Simon's theory of bounded rationality),[2] radical disaster scholars emphasized that disasters resulted from uneven political economic relations which made certain people and places more vulnerable than others to hazard impacts (O'Keefe *et al.* 1976). Political ecology emerged out of these critical studies on vulnerability. Broadly speaking, political ecology directs attention to the way uneven power relations shape environmental outcomes. While formative work echoed Marxist historical-materialist themes (Blaikie and Brookfield 1987), later iterations have folded in Foucauldian-inspired discourse analysis and governmentality studies (Peet and Watts 1993) and Latourian Actor-Network Theory (see Castree 2004).

Political ecology's focus on politicized human–environment relations directs attention to the effects of uneven power relations: how they limit access to natural resources and influence vulnerability and adaptive capacity (Blaikie and Brookfield 1987; Pelling 1998), are expressed through multi-scalar chains of causation (Bryant and Bailey 1997) and shape the possibilities for knowing and engaging with the non-human (Braun 2002; Robbins 2004). Accordingly, a political-ecology sensibility has been essential to many critical analyses of resilience initiatives. In particular, it has enabled scholars to examine how the dissemination of expert knowledge on hazards, risk and vulnerability has become a key vector of uneven power relations in the global political economy (Cretney 2014; Grove 2013a, 2014a; Walsh-Dilley and Wolford 2015).

Political ecology thus zeroes in on an apparent will to truth that underlies resilience programming: the tendency for resilience to fold progressively wider swaths of social and ecological life into a calculative neoliberal governmental rationality. From this perspective, resilience approaches do not offer a neutral, objective, and value-free account of social and ecological dynamics, but instead are bound up in the ongoing neoliberalization of human–environment relations. However, as we will see, this mode of critique has difficulty accounting for resilience theory's designerly elements premised on a will to design rather than a will to truth.

Resilience scholars' practices of critique extend this spirit of transgression: their critical engagements with command and control demonstrate that environmental science and management practices founded on modernist rationality, its prescriptions of optimization and efficiency, its utilitarian worldview, and its underlying liberal will to truth no longer hold purchase on complex social and ecological processes. For example, Fikret Berkes and Carl Folke (1998: 1) argue in language critical scholars would recognize that command-and-control management strategies have their roots in "the utilitarian and exploitative worldview which assumes that humans have dominion over nature… [and that] convert[s] the world's life-support systems into mere commodities." For Carl Folke and colleagues (2007), these resource management practices view nature only as a store of raw materials whose value can only be realized through generating wealth. And C.S. Holling and Greg Meffe (1996) emphasize that command-and-control strategies distort the possible ways humans might interact with nature, reducing this relation to questions of economically efficient resource production.

These arguments fit easily within political ecology – and critical debates on human–environment relations more broadly (Robbins 2004; Whatmore 2002). But ecological resilience theory takes these critical insights in different, designerly directions. In particular, resilience scholars' melding of designerly aesthetics with functionalist sociology and Simonian behavioral science means that they are not concerned with uncovering how power shapes environmental management practices. As we saw in Chapters 2 and 4, from a functionalist perspective, power relations and their effects are problems institutional design can accommodate (Lebel *et al.* 2006; Folke *et al.* 2007). Indeed, Per Olsson and colleagues (2014: unpag.) assert that techniques such as co-management, adaptive governance and transformation allow scholars to address "power sharing, decentralization and devolution of management rights, power asymmetries, and injustices that arise from structural inequalities of power." And much resilience work emphasizes the need to empower marginalized and vulnerable peoples. For example, a sizeable body of work has emerged around the topic of climate change adaptation, which focuses attention on the institutional, psychological and cultural constraints that limit individuals' adaptability (Brown and Westaway 2011; Eakin and Leurs 2006).

Thus, for resilience proponents, resilience offers a critical approach to environmental management because it attempts to *design out* power relations and their effects through experimental transformations. There is much to be said for these interventions, for they offer the promise of overcoming social and ecological marginalization in novel ways. But at the same time, these nominally critical maneuvers (or as we will see below, *designerly* critical maneuvers) also open up social and ecological life to novel forms of cybernetic regulation and control. In order to begin unpacking these tensions and juxtapositions, the next section examines the form of environmental power that circulates through designerly practice – and thus, resilience scholars' efforts to design out structural power effects.

THE POWER OF DESIGN

Critical research has detailed how, in practice, the empowerment resilience initiatives promise is often superficial and ineffective (Nadasdy 2003, 2007; Grove 2014a). In part, this results from the narrow visions of power resilience scholars mobilize to inform their understanding of how and why power relations affect sustainability and social and ecological justice. Resilience theory draws on an implicit modernist political imaginary, which views power as an entity that a sovereign individual can hold and exercise over others according to his will (and in this imaginary, the sovereign is always implicitly male). The exercise of power, in turn, involves actions that limit the ability of others to exercise their own will. Power is thus a negative entity: it is the capacity to prevent others from acting in accordance with their will. Politics then refers to the processes, techniques, and institutions that limit the arbitrary exercise of power. A modernist imaginary brackets power and politics from other domains of sociable and environmental life: understanding power and politics falls to the disciplines of political science and international relations; it does not bleed through these disciplinary borders into other fields such as economics, sociology, culture, or environment (see Walker 1993).

Resilience scholars' efforts to address "power sharing, decentralization and devolution of management rights, power asymmetries, and injustices that arise from structural inequalities of power" (Olsson *et al.* 2014: unpag.) thus arise out of a modernist political imaginary that views power as a problem of oppression, and politics as a technical pursuit to design and implement the most effective capacity-building initiatives (Grove 2014a). While these are certainly noteworthy goals, their underlying political imaginary blinds analysts to the specific forms of power that operate through transformative initiatives. As Michel Foucault (1990) has demonstrated, empowerment and freedom construct their own power relations and systems of subjection. The promise to empower often does little more than fold human and non-human entities into calculative programs of governmental regulation and control.

Foucault's concept of *biopower* provides one way to account for power outside the modernist problem of oppression and domination. In his 1975–1976 Collège de France lectures, 'Society Must Be Defended,' he characterizes biopower as a form of power that "makes live" and "lets die." That is, biopower operates through seemingly banal and apolitical efforts to promote the security, development, and well-being of individuals and collectives. He contrasts this with sovereign power, the power to "make die" and "let live." If sovereign power is the negative power of appropriation that involves techniques such as taxation, imprisonment or execution (appropriating money, mobility and life, respectively), then biopower is a positive power of security that operates through techniques such as education, statistics, and actuarial calculations (securing a population's continued circulation and growth in an uncertain milieu; Grove 2017).

Foucault proposed biopower as an alternative to the narrow and confining modernist political imaginary. Rather than objectifying and reifying power – even as an individual's positive "power to" accomplish something (Olsson *et al.* 2014) – biopower offer a model of power that focuses on what Maurizio Lazzarato (2006: 12) calls "a multitude of forces acting and reacting in relation to each other." From this perspective, power is not a thing that can be held by an individual, but is rather a *relation* that exists only as it is enacted through interactions. It is neither negative nor positive. Instead, it is *productive*: power produces both objects of power, such as abnormal human and non-human others requiring regulation, and subjects with specific kinds of desires, interests and capacities. In other words, the sovereign individual with interests is an effect of power relations that need to be analyzed and explained as such, rather than a coherent category and foundation of political analysis. Analyzing the operation of power relations, in turn, requires a *topological sensibility* attuned to the ways different techniques of biopower and sovereign power (such as statistics, observations, measurements and evaluations, educational training, physical punishment, actuarial calculation, or – as we will see – design) are combined and re-combined with one another in response to specific problems of government (Collier 2009). Here, government does not refer to formal political institutions, but rather the strategic practice of regulating human and non-human others – the "*conduct of conduct*," in Foucault's (2007a) phrasing.

Biopower thus directs attention to the calculative and strategic dimensions of power: the set of objectives, or visions of proper conduct, and mechanisms deployed to realize these objectives. As such, biopower is not a deductive, explanatory category; it is a provocation to inductively examine the contextually specific ways power attempts and fails to regulate life itself (Dillon and Lobo-Guerrero 2008). This means that strategies and techniques of biopower refer back to specific visions of life that need to be secured or developed – whether this "life" is thought as an individual, a biological population situated in an uncertain future, or a complex social and ecological system (Anderson 2012; Grove 2013c).

For example, Foucault (1995) details a number of disciplinary techniques that emerged in the 1700s and 1800s designed to regulate life understood as individuals with specific bodily capacities. Individualization, monitoring, and education enabled these capacities to be brought in line with ideal forms of behavior, whether this involved military procedures such as breaking down, cleaning, and re-assembling a rifle, penal practices designed to produce docile incarcerated subjects, or educational techniques that sought to train youth for wage labor. Similarly, biopolitical techniques such as statistics and demographic calculations enabled the state to visualize its population and resources as coherent entities whose productive potential could be maximized through economic, social, and public health initiatives designed to increase the health and productivity of a people and territory (Foucault 2007a). Security techniques such as risk calculations enabled individuals to visualize potential outcomes from actions taken in the present, and thus chart a course for safety and security in an uncertain milieu (Ewald 1991; O'Malley 2004). In each case, the exercise of power relations is intimately connected with the scientific production of certain forms of truth about "life" as such. Biopolitical techniques such as statistics involved the identification a[…] of specialized knowledge about distinct spheres of existen[…] population (demography and biology), the economy (eco[…] ture (anthropology) and society (sociology; see Dean 20[…] security techniques relied on the development of risk manag[…] calculations connected with the actuarial sciences (Hacking 1990).

A biopolitical imaginary thus suggests that new experiences of spatial interconnection and temporal emergence that gave rise to the complexity sciences will also usher in new techniques of regulation and control. Design, cybernetics and, later, resilience theory mobilize a distinct form of biopower, what Brian Massumi (2009) calls "environmental power," to visualize and manage complex social and ecological systems (see Box 6.2). As the previous chapter detailed, the term "environment" here refers to a system comprising distinct components functionally combined with one another. Environmental power refers to a form of biopower that draws elements into functional, systemic arrangements. In other words, it is the power of design: the power to render concrete, contextualized and embodied material entities as rational abstractions, imaginatively visualize how these abstractions might be functionally synthesized with one another, and synthesize these abstractions into artifacts – provisional solutions to indeterminate problems of complexity. Environmental power is thus the power to align an interior and exterior in the absence of transcendental determinants of meaning and value. In other words, it is the power to design systemic effects, which operates on the functional *potential* that inheres in any rational abstraction in order to create certain possibilities for individual and collective existence.

Box 6.2 ENVIRONMENTAL POWER AND AFFECTIVE ENGINEERING

A key distinction between environmental power and other techniques of biopower is the centrality of *affective engineering*. Here, the term *affect* signals the transpersonal, pre-individual capacities of bodies to affect and be affected (Massumi 2002). Affects can be thought as a kind of atmosphere that surrounds bodies and charges their appetites – we are able to think, feel, and interact with the world around us in the way we do because of the way we affect and are affected by those bodies that make up our surroundings (Anderson 2009; Grove 2014a). Affects are abstract – in Deleuze and Guattari's terminology, they are virtual, or "real without being actual" (Massumi 1992). Affect signals a constitutive openness to the world: as affective beings, we are constituted through immeasurable relations with our surroundings. This means that our desires, appetites and interests do not arise from a sovereign will, but instead emerge as our surroundings affect us.

As Nigel Thrift (2004) has demonstrated, contemporary governance strategies often involve techniques of affective engineering designed to shape how we think, feel, desire and act. Advertising campaigns are one of the more blatant examples here, but we can also recognize affective engineering in the US government's efforts during the early 2000s to modulate citizens' fears and anxieties through color-coded "terror alert" status updates (Massumi 2005) or Simonian-influenced strategies of paternalistic governance designed to operate on individuals' (affective) choice environments (Jones *et al.* 2013; Chandler 2013). Peter Adey (2008) has also traced how environmental power operates in the practice of airport design to produce affects such as circulation, mobility and confidence. The color and placement of airport signage can have certain beneficial (or detrimental) systemic effects: certain colors, such as yellow or green, attract attention, while others, such as red, increase anxiety. In the busy, hectic and stressful environment of airport travel, signs designed with the former colors can help ensure travelers recognize where to go as they move through the airport, and lessen their anxieties as they move about. The functional qualities of different colors thus help produce a complex system – the functional "environment" of the airport – that produces a specific service: the mobility of people and goods across the global economy.

Environmental power is thus the form of power exercised through design. Design theorists themselves refer to this process of abstraction and creative, imaginative functional re/combination with the provocative term "placement." First elaborated by design theorist Richard Buchanan (1992), a placement can be thought as any device that enables a designer to de-contextualize an object, render it an abstraction, and explore the possibilities that might result when that object is placed within other contexts. As he writes,

> [placements] allow the designer to position and reposition the problems and issues at hand. Placements are the tools by which a designer intuitively or deliberately shapes a design situation, identifying the views of all participants, the issues which concern them and the invention that will serve as a working hypothesis for exploration and development.
>
> (Buchanan 1992: 17)

Here, we can see the influence of Simon's bounded rationality. If human cognitive capacities are unable to fully grasp the entirety of complex life, then the designer needs to invent ways to visualize problems from necessarily limited perspectives of different stakeholders. Each stakeholder will offer a potentially useful slant on the problem, but it is the designer's task to create ways of synthesizing these perspectives in order to develop a novel understanding of the problem at hand and potential solutions – or, as Buchanan puts it here, the "working hypothesis for exploration and development."

We can thus think of placement as a device for moving from one bounded rationality to another: the problem gets "placed" in different contexts, different situations, and these "placements" help the designer see aspects of the problem she otherwise might have missed. Design studies scholar Barry Wylant describes Buchanan's sense of placement in telling language:

> the playful manner in which this [contextualization] occurs is reflected in Buchanan's terminology, where the act of contextualizing something is referred to as a "placement." Here something is contextualized without a prior sense of its dominance or importance. Context is considered temporarily, without commitment, as a means of learning about any issues on the table in a particular problem.
>
> (Wylant 2010: 288)

Buchanan's theory of placements thus transforms the critical significance of context. Context no longer simply demonstrates the partiality and contingency of universalizing truth claims. In the hands of designers, it provides a device to generate new meanings for an object, and design signals the process of playfully exploring different contextualizations – or placements – in order to produce new meanings and outcomes.

While the term "placement" is specific to design theory, it signals a process of abstraction and functional re/combination that occurs in a number of fields shaped by designerly aesthetics. Resilience-based approaches to transformation offer a case in point. For resilience scholars, transformability is a process of "defining and creating novel system configurations by introducing new components and ways of governing SESs [social and ecological systems], thereby changing the state variables, and often the scales of key cycles, that define the system" (Olsson *et al.* 2006: unpag.; see also Chapter 4). The logic here is that altering system configurations in some way (for example, through creating a new institution to manage stakeholder conflict) will give a system's components new identities and functions, which will enable the system to operate in a new way. In designerly terms, this is a question of how, given a specific arrangement of elements such as organizations, interest groups, ecological processes, legal frameworks, and so forth, resilience initiatives might create new functional "environments" that enhance beneficial adaptive capacities while reducing rigidity and/or undesirable resilience. Altering system configurations involves re-purposing existing artifacts – existing governance regimes – and giving them new identity, meaning and function. Of course, the precise effects of these interventions are indeterminate and cannot be known in advance: "the transition to adaptive governance can only be navigated, not planned" (Olsson *et al.* 2006: unpag.). There is no transcendent determinant of proper governance; an effective governance arrangement can only be worked out through the adaptive, stepwise process of designing targeted and bounded interventions, monitoring their effects, and adjusting them as needed (see Box 6.3).

Box 6.3 DESIGNING TRANSFORMATIONS

The process of managing social and ecological transformations is shot through with designerly sensibilities. Resilience approaches disaggregate conflictual processes of social and ecological change into Simonian, hierarchically organized functions that can be creatively re/synthesized with one another. For example, resilience scholars disaggregate transformation into distinct functions that include preparing a system for change, the role of leadership, and the importance of windows of opportunity that crises offer (Olsson *et al.* 2006). Each provides scholars with a series of qualitative and, in some cases, quantitative variables they can use to visualize and manage the process of transitioning to more inclusive adaptive governance arrangements. For example, *preparing a system for change* involves activities that build social capital by enhancing both vertical ties (relations between community members and state and non-governmental organizations) and horizontal ties (relations between community members themselves). These ties facilitate knowledge-sharing and integration. They also aid in the development of "shadow networks," or informal governance arrangements that can advance alternative management approaches. Similarly, *leadership* can introduce new vision, create new meanings and values and promote new alliances between disparate groups (Folke *et al.* 2005; Westley 1995). These factors all enhance the capacity for groups to collaborate on problem identification and solution in novel ways. Finally, *windows of opportunity* are crises that can trigger systemic change, provided the system has been prepared beforehand. Crises can be artificially provoked, such as environmental activist lawsuits that force environmental management changes, or they can follow from unpredictable events in the wider environment.

The resilience scholar's goal, much like the Simonian administrator (Simon 1997), is to combine these functional levels in ways that facilitate rational – adaptive – governance in a complex world. A system that has been prepared for change, often through visionary leadership, will be able to recognize a window of opportunity as it arises and transition to a more sustainable and just environmental governance regime.

Thus, just as environmental power creates worlds that facilitate certain forms of sociability rather than others (see Box 6.2), so too does resilience create socio-ecological worlds that facilitate certain forms of human–environment relations. Resilience techniques exercise environmental power through producing truths about social and ecological phenomena as rational abstractions. These techniques transform a world of indeterminate affective socio-ecological potential into a series of determinate possibilities: socio-ecological relations can be either adaptive or maladaptive (Grove 2017; see also Deleuze 1995a). At the same time, these techniques also enable novel practices of regulation and control: the designerly manipulation of affective relations to enhance a system's adaptive capacities.[3] Kai Lee (1999: unpag.) signals this form of control in a provocative description of adaptive management:

> Adaptive management recapitulates the promise that Francis Bacon articulated four centuries ago: to control nature one must understand her [sic]. Only now, what we wish to control is not the natural world but a mixed system in which humans play a large, sometimes dominant role. Adaptive management is therefore experimentation that affects social arrangements and how people live their lives.

The parallel he draws is telling: if Bacon's efforts to control nature hinged on experimental scientific practices that produced objective knowledge, then the efforts of resilience scholars to control complex social and ecological systems likewise depend on experimentation. But this is a designerly form of affective experimentation that strives to create new systemic configurations through playfully recombining and repurposing existing systemic elements. As a form of "experimentation that affects social arrangements and how people live their lives," adaptive management recalibrates modern science's promise of control within a complex and indeterminate world (ibid.: unpag.). As a vector of biopolitical control, resilience initiatives configure techniques of environmental power with other techniques of disciplinary, biopolitical, security, and sovereign power to produce environmental effects: that is, to *design* social and ecological systems that function in certain ways rather than others. As such, resilience names one particular set of techniques for intervening in an indeterminate world – and it opens a field of ethical and political struggle over what kinds of possibilities these functional re/combinations create, and what potentialities they foreclose or negate. To illustrate these ethical and political complexities, the next section reviews a case study of community-based disaster resilience in rural Jamaica.

ENVIRONMENTAL POWER IN ACTION: DESIGNING CULTURES OF SAFETY IN RURAL JAMAICA

In 2005, the United Nations' Hyogo Framework for Action made "building the resilience of nations and communities" a key pillar of international disaster management practice (UN 2005). In distinction to the technocratic focus of the preceding International Decade for Natural Disaster Reduction (IDNDR; see UN 1989), the Hyogo Framework emphasized the need to build resilient "cultures of safety" through community-based disaster management initiatives (Grove 2013a). Since that time, community-based disaster management (CBDM) has become a fixture in disaster management and development projects (Maskrey 2011). While initially connected to radical programs of participatory consciousness-raising and radical political economic change, most CBDM projects now focus on more immediate concerns of reducing human suffering from disasters and environmental change (Wisner 1993). Without doubt, the operationalization and widespread adoption of participatory approaches into CBDM programming has reduced disaster related fatalities – an important achievement in its own right. But it also creates a vector for new forms of power relations to transform social and ecological relations. Humanitarian efforts to secure life through CBDM involve a series of educational and participatory techniques that further render life itself – the affective relations between humans and their environments – objects of governmental control. These techniques enact an immunological politics that turns life against its own vital force (Grove 2014b; Esposito 2011). In short, community-based adaptation and resilience programming offers another way of visualizing, operationalizing, controlling, and in the last instance, negating the population's constitutive adaptive capacity, whose potential to generate socio-political difference threatens fragile state-based orders throughout the developing world.

Fieldwork from my collaborative research with Jamaica's national disaster management agency, the Office of Disaster Preparedness and Emergency Management (ODPEM), will help illustrate this argument (see Grove 2013b, 2014a, 2014b). I assisted with project monitoring on two donor-funded CBDM initiatives: the Tropical Storm Gustav Recovery Project (or the DFID project), and the Building Disaster Resilient Communities Project (or the CIDA project). Both projects involved a form of environmental engineering designed to draw together communities, individuals, their surroundings, local and expert forms of knowledge, state and non-state agencies, and desires for security and well-being into systemic assemblages that produce resilient, self-organizing communities capable of adapting in ways that reinforce rather than challenge Jamaica's fragile neoliberal order. They attempt to produce these systemic effects through two specific techniques: participatory education activities, and individual and focus group interviews. While each of these techniques has a long history in disaster management, we will see that their deployment in CBDM initiatives transforms them in relation to the designerly imperative to create and functionally synthesize rational abstractions. Each is considered in turn.

Abstracting local knowledge through education programming

Education has long been a fixture in emergency and disaster management programming. For example, in Cold War civil defense planning, education activities informed citizens and communities about the nearest nuclear fallout shelters, proper procedures to follow during an emergency, and how best to prepare themselves for a nuclear attack. While this disciplinary function is still an important part of disaster preparedness, the recent attention practitioners and academics have given to pre-event risk mitigation and resilience transforms the meaning and effects of education. This transformation reflects a new problematization of local knowledge. Education no longer addresses community members' lack of knowledge on hazards and vulnerabilities; indeed, resilience approaches now recognize that local people possess their own forms of knowledge, which are often more effective at reducing vulnerability than expert knowledge. Instead, the problem is that this local knowledge has developed autonomously from the institutions of capital and state rule. ODPEM's 1999 National Hazard Mitigation Policy sums up this problem succinctly when the authors note that traditional knowledge derives from "a history of autonomously adapting to hazard events and changing environmental conditions" (ODPEM 1999: 5). This local knowledge potentially fuels alternative practices of resilience, which might subvert the goals of disaster resilience and climate change adaptation (Grove 2013b).

Against this threat, educational activities now attempt to abstract contextualized, local experiences of hazards and vulnerability – which may explain these conditions in terms of political economic inequality or fatalism – and re-combine them within the discourses and categories of expert knowledge. For instance, the Jamaican Social Investment Fund, which frequently partners with ODPEM on CBDM, utilizes a computer simulation to demonstrate the impacts of hazard events, such as floods and landslides, on participants' communities. These instruments each recode everyday experiences through the concepts of geology, hazards studies, and climate change science. As I have detailed elsewhere (Grove 2014a), this recoding operates on both cognitive and affective levels.

In de-contextualizing local knowledge, education activities attempt to render this knowledge functionally abstract, and thus capable of affecting and being affected by other systemic components in new ways. How does this abstraction and functional re/combination occur? Training videos, computer simulations, and other participatory education instruments attempt to dissociate local knowledge from the longer history of 'autonomously adapting' to vulnerabilities, as above, and specify it through the categories of expert knowledge. In the process, the environment becomes inscribed as a source of fear and insecurity: a hillside becomes a potential landslide; rain becomes a potential flood. This is a specific kind of insecurity that can be handled through disaster management programming: the environment is something community members should fear, not because centuries of colonial and post-colonial rule have left their bodies vulnerable to hazard impacts, but simply because environmental change is increasing their hazard risks. As Chapter 2 noted, the latter is a functionalist form of vulnerability that can be managed through participating in CBDM activities and learning how to properly prepare for and respond to disaster events. Educational techniques thus attempt to manipulate the affective relations between people and their biophysical surroundings in order to construct relations of fear and unease that will compel people to take part in participatory CBDM and utilize the knowledge gained there.

Functionally synthesizing resilient communities: interviews and transect walks

Second, interviews and focus group activities engage in a similar process of affective engineering. The interviews and focus groups I conducted with community members participating in the DFID and CIDA projects subtly manipulated affective relations such as fear, hope, mobility, health, and hunger to manufacture resilient communities. As above, some questions encouraged people to think about their surrounding hillsides, rivers, riverbanks, and trees in terms of potential landslides, floods, blocked roads, and other hazardous phenomena. Other questions sought to identify residents with specific kinds of knowledge, skills, and resources who could assist others after a disaster event. Some of these valued characteristics included people with medical and first aid training, people with large houses and secure roofs for shelter, people with chainsaws to clear blocked roads, and shopkeepers willing to extend credit to community members. If individuals were unaware of residents with these skills and resources, we would then share information we had gathered so far.

Our interviews and focus groups thus attempted to abstract existing relationships between neighbors and friends (and possibly enemies!) and recompose them in terms of these affective capacities they could provide. Affects such as mobility are constructed as people identify and call on residents with chainsaws to help clear roads. Affects such as health are constructed as people identify and call on neighbors with medical, first aid, and search and rescue training. Affects such as sustenance are constructed as people identify and call on local shopkeepers for relief supplies. In other words, CBDM techniques abstract and functionally re/combine community relations into potential resources that residents can draw on in the aftermath of a disaster. These affective capacities – capacities to *become* fed, mobile, healthy, sheltered and so forth – enable communities to draw on each other in the aftermath of a disaster, and thus become active agents in the disaster response process.

The indeterminate politics of CBDM

These two examples demonstrate how CBDM attempts to de-territorialize local knowledge on hazards, risks, and vulnerability, and re-territorialize it within a diagrammatic series of relations between community members, their surroundings, state agencies and their partners, expert forms of knowledge on hazards and adaptation, and international donor organizations. To the extent that educational and participatory techniques successfully engineer affective relations such as fear, hope, mobility, health, and sustenance, CBDM designs resilient individuals: subjects with the knowledge, capacities and confidence to respond to disaster and adapt to change in "proper" ways that do not undermine state order. Collectively, these resilient individuals comprise resilient communities: self-organizing systems that spontaneously adapt to surprises in the appropriate manner, without the need for state intervention and control.

In short, CBDM operates through techniques of environmental power that abstract and functionally re/combine community relations in ways designed to produce resilient cultures of safety. While many critics dismiss these efforts as simply furthering the neoliberalization of socio-ecological relations – for good reason, as we will soon see – these techniques are *politically polyvalent*. This means they can produce multiple, divergent political effects, depending on how they are deployed in specific situations in response to specific problems of indeterminacy. The Jamaican case is exemplary here. On one level, CBDM produces biopolitical effects that immunize Jamaica's neoliberal order against the threat marginalized people's radical adaptive capacity poses to this order. This is an axis of conflict that stretches back to the colonial era, when emancipation created a politically and economically surplus population of freed slaves excluded from the series of rights and responsibilities of liberal rule. While this left them exposed to the arbitrary violence of colonialism's race-based order (Thomas 2011), their forms of association were also not delimited by a liberal political imaginary (Duffield 2007, 2010). Since emancipation, a number of techniques have been deployed in various combinations to govern this excess potential, such as sovereign violence exercised on black bodies, and disciplinary and biopolitical programs of education, improvement and development (Meeks 2000; Bogues 2002; Scott 2003).

CBDM is the latest fold in this four hundred-year struggle. It institutes an *immunological* relation to this excess potential, which positions adaptive capacity as both the source of threat and security in an indeterminate world. As political theorist Roberto Esposito (2011) argues, in biopolitical terms immunization seeks to protect qualified life against a threatening other. It involves a process of introjecting the other into the self: that is, it attempts to protect the self against the other by introducing a controlled form of the other into the self. Immunization thus alters both the self and other in the name of protecting the self. For Esposito, this other is typically an extensive body, such as a terrorist or immigrant. But the environmental techniques described above turn this other into an intensive force: marginalized populations' immanent adaptive capacities that result from affective relations between people and their surroundings. Because adaptive capacity is external to the state and donor agencies, there is always the possibility, however thin, that people might adapt to environmental change in ways that threaten state-based neoliberal order. To inoculate the state against the threat of resistance in the aftermath of disaster, CBDM attempts to introject the populations' inherent radical adaptive capacity in an artificial and controlled form – the capacity for proper forms of self-organized disaster response outlined above. Recoding the relations between community members, their surroundings, and state agencies designs an artificial form of the population's radical adaptive capacity. This is a form of adaptive capacity proper to neoliberal order, one that can be governed and regulated through CBDM: the adaptive capacity held by the resilient and disempowered individual in need of participatory programming (Grove 2014a).

However, on another level, CBDM gives individuals and communities in rural Jamaica the ability to reduce suffering and manage catastrophes without relying on assistance from state agencies or international donors. In the context of global austerity budgeting, these capacities are often the difference between life and death for many afflicted peoples. Moreover, in the context of Jamaican politics, the empowerment CBDM offers stands in opposition to patron-clientelism, a specific form of political practice in Jamaica that operates through a series of dependencies and obligations (Stone 1983). The roots of patron-clientelism lie in the plantation system and the material and psychic dependencies plantation owners created amongst workers, who came to rely on this oppressive system for their own well-being (Beckford 1972). This dependency carried over into formal democratic politics: participation in the two-party electoral system is based on popular expectations for material benefit. Party elites use the state to distribute resources to loyal constituents, often brokered through neighborhood leaders or "dons." In this system, disaster recovery has traditionally been another spigot for elites to control. CBDM in Jamaica implicitly confronts these dependencies (Grove 2013b). Education programming attempts to give community members different forms of knowledge on their social and environmental vulnerabilities, with the hope that they can begin identifying new areas of hazard exposure themselves. Additionally, grant-writing training programs attempt to give community leaders the skills to identify and secure funding for vulnerability-reduction activities from national and international donors themselves, rather than relying on state resources. While both of these projects can clearly be read as strategies of neoliberal responsibilization, in the context of Jamaican clientelism, they *also* represent an important departure from the entrenched dependence of the urban and rural poor on party and state elites.

Community-based disaster resilience is thus politically indeterminate: at one and the same time, it is a vector for challenging the clientelist legacy of Jamaican politics, *and* reinforcing the ongoing neoliberalization of social-ecological relations. This indeterminacy returns us to the paradoxical relation between resilience theory and power: at one and the same time, resilience approaches reinforce unequal neo-colonial political economic relations, even as they strive to create new possibilities for ordering state–society–ecology relations in a complex and uncertain environment.

To explore how resilience theory holds these divergent tensions together, the next section returns to design studies and the unique style of designerly critique that animates resilience initiatives.

RESILIENCE THEORY AND DESIGNERLY CRITIQUE

As we have seen, resilience thinkers engage in at least two distinct critical maneuvers. *First*, as Chapter 4 detailed, they engage in a style of critique that demonstrates the limits of command and control, which serves as the basis for their subsequent calls for alternative strategies of adaptive management, adaptive governance, and other techniques of complexity that can design-out power effects. But there is a second critical maneuver as well: the *act of fabricating* systemic capacities through playful re-contextualization. This is *designerly critique*: the practice and process of creating artifice in an indeterminate world. In the case of resilience planning, this often involves adaptively designing institutions that can "properly" align governance mechanisms with the complex social and ecological systems they seek to control (Folke *et al.* 2007). Barry Wylant brings these two forms of critique into sharp relief when he suggests that,

> [the] postmodern inclusion of historical context or socio-cultural aspects such as feminism contribute to useful bodies of theory, yet do not necessarily inform how such information can be considered, implemented and used in the creative act. Application requires an understanding of which elements are essential or critical in making the new thing. In this, a fundamental question as to where the *critical* resides emerges an overriding challenge of modernity in all its forms.
> (Wylant 2010: 227; emphasis in original)

Here, Wylant differentiates conventional forms of critique – the "postmodern inclusion of historical context or socio-cultural aspects such as feminism" that demonstrate the present's contingency and partiality – from a designerly critique that involves the practice and process of creating artifice in an indeterminate world. In his reading, conventional critique (hereafter simply "critique") and designerly critique diverge at their relation to partiality and the limits of knowledge. If critique relies on strategies of contextualization to undermine totalizing knowledge claims, and show that all knowledge is partial and situated, designerly critique departs from the recognition that all knowledge is partial and situated, and asks *what we can do* with this limited knowledge.

To be sure, Wylant's description mobilizes elements of pragmatism, such as his imperative to "apply" and "use" partial and bounded knowledge. For Bruno Latour (2004), this form of critique replies to "matters of concern" rather than "matters of fact": the imperative to "make the new thing," as Wylant puts it, emerges out of specific, intractable problems that draw diverse people and things together (see also Marres 2012). But in the remainder of this chapter, I want to dwell on how designerly critique recalibrates Kant's critical sensibility within a novel spatio-temporal context. Wylant's vision of designerly critique is *not* set within and against a world trapped in the vise-grip of a totalizing will to truth. As we saw in the previous chapter, a complex world is already destabilized. Truth is always partial, contingent and bounded (Simon 1955), categories of knowledge lose their permanence and coherence, and control no longer hinges on prediction. This is an experience of a world grasped through the sensorial regime of modernist design aesthetics. In a world grasped as a functional environment, where "what exists" is rational abstractions functionally synthesized with one another, totalization is impossible. Functionality – the essence of truth and beauty after Bauhaus – is relationally constituted, and thus anathema to totalizing, predictive knowledge. What matters is less what something *is* than what it *does* – and what it *does* often hinges on how something can be creatively *synthesized* with other things.

Thus, designerly critique does not confront totalization. It does not attempt to transgress the present arrangement of things in order to free the potential to create different arrangements of things, *because this potential already saturates the present*. A rational abstraction is by its nature virtual: it is pure abstraction, potentiality waiting to be actualized through the *nouveau*-critical act of synthesis (Wark 2004). Designerly critique thus asks *what something might become*: what new things can be made from the existing arrangement of things? In short, the challenge for critique is no longer to generate difference, but rather to make difference become useful in particular ways.

Elizabeth Grosz's (2009) engagements with Deleuzian philosophy help us specify how designerly critique operates in philosophical terms. In her Deleuzian terminology, designerly critique merges art and science. Art refers to practices that connect us with the experience of otherness, and thus signals the partiality of our own existence and the potential for the world to be other than it appears to us. It signals practices of *becoming-otherwise*. Science, in turn, refers to practices that abstract singular existents in ways that enable them to be folded into strategic calculations and interventions. Science makes what exists useful to specific needs. It thus signals practices of *becoming-useful*. In these terms, designerly critique blends techniques of art and science in ways that allow *becoming-otherwise* to *become useful*.[4] This designerly recalibration of critique means that designerly critique does not *transgress*; instead, it *makes transgression useful* for specific problems.

The end result is that resilience thinking transvalues core elements of critical thought (Chandler 2014a). That is, it gives new meaning and significance to key expressions of transgressions that destabilize a will to truth. For our purposes here, two transvaluations are essential to understand how resilience transforms critique into a mechanism of cybernetic control (although others are surely possible). First, it approaches contestation and conflict as opportunities for social learning, rather than signs of uneven power relations that betray the objective neutrality of a given management intervention. This results in research focused on the barriers to transformation and the need to overcome resilience traps. Second, it treats alternative forms of knowledge on social and ecological dynamics as potentially useful forms of bounded knowledge that can be synthesized with expert scientific and bureaucratic knowledge, rather than forms of "subjugated knowledge" (Foucault 2003) that reveal ongoing social warfare. This results in research focused on identifying distinct knowledge systems and exploring the possibilities for synthesizing these systems. In both cases, expressions of social and ecological difference become the motor rather than saboteur of adaptive and reflexive resilience interventions. In other words, becoming-otherwise becomes useful.

The following sections unpack how resilience thinking transvalues conflict and subjugated knowledge. We turn first to the category of conflict.

PRAGMATISM, PUBLIC CHOICE THEORY, AND THE NATURALIZATION OF CONFLICT

Kai Lee (1993) was one of the first resilience proponents to suggest that conflict could be a stabilizing force if handled appropriately. His influential adaptive management text, *Compass and Gyroscope*, uses "gyroscope" as a metaphor for an instrument that produces stability and certainty *out of* instability. In basic terms, a gyroscope is a device that redistributes destabilizing motion in a way that isolates a stable interior and insulates it from potential disruption. In other words, it uses instability to create stability. If "compass" names the stepwise, adaptive search for satisificing solutions to complex problems (see Chapter 5), then "gyroscope" names the institutional forms – artifacts – that allow inevitable conflicts that emerge during this search to become beneficial to this search process. Indeed, Lee (1993: 10) warns of the dangers "unbounded conflict" poses to sustainability: it can "destro[y] the long-term cooperation that is essential to sustainability." However, "like a spinning gyroscope, competition is motion that can stabilize" (ibid.). But this raises the problem of how to bound conflict, for political competition requires unwritten and written rules, or institutions that create the context in which political action takes place. This is inherently a design problem, for it involves the creation of artifice – institutions – within an indeterminate and complex environment that undermines predictive certainty.

This question of designing institutions in a complex world has a long history in public choice theory, which we will explore shortly. But Lee's particular slant on the problem helps us clarify how resilience approaches visualize and value conflict. Lee turns to John Dewey's pragmatism, and specifically Dewey's efforts to reconceptualize democratic governance within the context of the US's burgeoning industrialization in the early twentieth century (see especially Dewey 1927). In brief, as Lee notes, Dewey suggested that industrialization's tendency to socialize production had rendered individuals unable to recognize the complex relations and processes that meet their daily needs.[5] This is a precursor to Herbert Simon's thought on complexity and bounded rationality (Crowther-Heyck 2005), but Dewey focuses on the problems complexity poses to democratic governance. Specifically, the liberal-democratic ideal places its faith in individual (political) choice, but in a complex world, their decisions will unavoidably draw on partial and inaccurate understandings of their situation. If individual knowledge is an inadequate ground for democratic governance, then democracy, Dewey argues, should be reformulated into a process of collective problem-identification. In a complex world, democracy involves an ongoing reflexive process that identifies problems, forms affected publics, and designates scientific and governmental experts to manage these problems.

Even though Dewey gave centralized expertise a privileged role in *solving* problems, he still emphasized the role collective learning plays in democracy more generally. Specifically, the affected publics democratic institutions serve emerge as individuals reflect on common conditions and experiences and gradually learn about complex interdependencies that structure everyday life. Learning thus drives the process of democratic governance. Lee borrows this element of learning from Dewey's political theory. He uses the concept of "social learning" to designate how collectives can democratically engage with complex problems of sustainability that involve several competing factions, such as his case study of salmon conservation in the Pacific Northwest. Read through Dewey's pragmatic political philosophy, Lee treats political contestation as a key mechanism for social learning: "political conflict can provide ways to recognize errors, complementing and reinforcing the self-conscious learning of adaptive management" (Lee 1993: 87).

Lee thus transvalues conflict through the category of social learning. Conflict is no longer a destabilizing force that reveals underlying power relations – and thus the contingency of the present. Instead, it becomes the means through which governance cybernetically identifies problems to address. In a complex world, conflict is unavoidable – *because individuals develop preferences based on partial and/or incorrect information*. The question is not whether conflict will arise, but how it arises. The precise form conflict takes will indicate specific design opportunities, or chances for managers to experiment with different organizational and institutional arrangements that can better align an interior (for Lee, competing interest groups) with an exterior (ecological conditions) in a complex and indeterminate environment. But of course, this hinges on designing institutions that enable conflict to generate rather than undermine sustainability.

This designerly sensibility towards conflict is a defining if often implicit feature of resilience scholars' work on adaptive governance and transformation. For example, Carl Folke and colleagues (2005: 448–449) contrast adaptive management with adaptive governance in language worth quoting at length:

> Although adaptive management focuses on understanding ecosystem dynamics and feeding ecological knowledge into management organizations, adaptive governance conveys multi-objective reality when handling conflicts among diverse stakeholders and, at the same time, adapts this social problem to resolve issues concerning dynamic ecosystems.

Lee (1999: unpag.) similarly suggests that, "the art of the conservationist [in conditions of competing interest groups] ... is to reconcile conflicting objectives, at least temporarily, so as to make agreements on land use and other protective actions possible." This process of adapting the "social problem" of competing views and "reconciling" conflicting interests is a synthetic process of making difference useful through re/designing institutions that govern resource use.

However, this problem of institutional design has a wide genealogy with roots in public choice theory. Importantly, this work provides a vision of the social as a product of artifice that informs resilience scholars' engagements with the problem of institutional fit in the 1990s (see Chapter 4). In brief, public choice theory is a branch of new institutional economics that emerged in the 1950s and 1960s. It sought to apply economic theory and methods to traditional problems of political science – particularly the provisioning of collectively consumed goods such as urban infrastructure services. In the main, this is a somewhat dry and predictable literature that offers market-based solutions to collective choice problems. However, at its margins, a handful of scholars developed sophisticated understandings of the artificiality of human existence. As Stephen Collier (2011, 2017) details, two public choice theorists in particular – James Buchanan and Vincent Ostrom – reformulated classical tenets of liberal political economic thought in response to the novel experiences of complexity Dewey and Simon articulated.[6] While both shared an aversion to centralized planning, neither blindly asserted that the market was the optimal mechanism for determining the provision of public goods. Although market logics might hold in certain situations, they recognized that it cannot account for the complex totality of human becoming. In their formulations, individuals choose certain behaviors for reasons that cannot be explained with recourse to utility maximization (e.g., Buchanan 1999; Ostrom 1976). Market rationality, like other transcendental determinants of value such as centralized state planning's social welfare calculations, could distort information on user preferences, and thus lead planners and economists to make flawed public good provisioning decisions.

Buchanan's and Ostrom's critiques of both the state and market express an underlying designerly sensibility and a cybernetic concern with efficiently organizing information flows within complex systems. For example, in their own ways, they both approach the provisioning of public goods as a design problem (see Collier 2017): the question of how to design artifacts (in this case, institutions) that will align an exterior (collectively consumed goods) with an interior (preferences of users) in conditions of indeterminacy (constantly changing user preferences). For example, writing with Gordon Tullock (Buchanan and Tullock 1962), Buchanan extended this designerly sensibility to the creation of constitutional order itself. Their concept of *constitutional choice* signals how humans create rules that structure everyday choices – including the constitution, or the rules that structure political order within a polity itself (see also Ostrom 1980). In their reading, there is no transcendent determinant that objectively guides decision-making on institutional rules, including at the constitutional level. Instead, constitutional choice involves a process of experimental policy changes within an indeterminate political economic context (see also Ostrom 1980). Buchanan suggests, in language resilience proponents will echo forty years later, that "the problem for the political economist is that of searching out and locating from among the whole set of possible [policy] combinations one which will prove acceptable to all parties" (Buchanan 1959: 131). This "searching out" function adapts Herbert Simon's (1997) earlier work on hierarchically arranged decision-making that does not strive for *the* optimal solution, but rather identifies *satisficing* outcomes. Indeed, Buchanan asserts that political economists should, "assist individuals, as citizens who ultimately control their own social order, in their *continuing search* for those rules of the political game that will best serve their purposes, whatever those may be" (Buchanan 1999: 250).

Elinor Ostrom's work on common-pool resource governance linked resilience scholars with public choice research on institutional design. In her early work, she blended Buchanan's theory of constitutional choice with Simon's ontology of hierarchically organized complexity to develop a conceptual model of "three worlds of action." Writing with Larry Kiser, she disaggregates institutional design into three levels with distinct rates of change: fast-changing operational rules, mid-range collective choice rules, and slow-moving constitutional rules (Kiser and Ostrom 1982). This framing formed the basis for her subsequent work on common-pool resource governance and later engagements with institutional design for sustainability. For example, her Nobel Prize-winning *Governing the Commons* (Ostrom 1990) directly confronted Hardin's (1968) "tragedy of the commons" thesis with an argument on the artifactual nature of resource governance institutions. As above, Hardin argued that individual self-interest and utility maximization will inevitably over-exploit common-pool resources such as fish or tree populations, and lead to the collapse of these resources without a market mechanism to rationalize individual decision-making. In contrast, Ostrom showed that communities had developed a variety of different institutional arrangements that successfully managed common-pool resources – in some cases, such as Netherlands water management, for centuries – *without relying on either the market or the state*. Instead, these institutional arrangements were artifacts, products of an indeterminate design process that she distilled down to eight abstract design principles (see Chapter 4, Box 4.3).

As C.S. Holling (2001: 398) recognized, Kiser and Ostrom's "three worlds" model of institutional change is an economic analogue to ecologists' understanding of panarchy. Ecological resilience theory and public choice theory converge around a common effort to recalibrate their disciplines – ecology and economics, respectively – around a cybernetic sensibility and its underlying designerly aesthetics. For example, writing in 1999, Elinor Ostrom suggests that "policy changes are experiments based on more or less informed expectations about potential outcomes and the distribution of these outcomes for participants across time and space" (Ostrom 1999: 520). In the same year, Kai Lee (1999: unpag.) echoes this language when he suggests in a quote referenced above, that, "adaptive management is therefore experimentation that affects social arrangements and how people live their lives." Both bodies of work share a common understanding of complexity based on Simonian hierarchy, and recognize the designerly qualities of institutional change. And importantly, both share an aversion to centralized planning *and* market-based sustainability solutions: there are no "panaceas" (Ostrom 2007) to problems of designing institutions for sustainability in a complex world.

The convergence between ecological resilience theory and public choice theory thus often takes on a critical appearance that challenges both state-based and neoliberal approaches to environmental management. However, the work of Buchanan, the Ostroms and others offers a distinctly *liberal* theory of change that inflects how resilience scholars view the social. As Collier's (2011) engagement with Buchanan's thought shows, public choice theorists view the social as artifice – an emergent effect of individual choices. Like other branches of liberal political economic thought, public choice theory is founded on an implicit ontology of individual choice: the world is made up of individuals choosing from among a range of different options based on their preferences. Of course, this is a *neo-liberal* theory:[7] public choice theorists spin a liberal ontology through a Simonian understanding of the boundedly rational, adaptive individual whose choices in an indeterminate world construct the institutional context that bounds their rationality. In this vision, conflict is an unavoidable fact of life as such: individual preference will necessarily be partial and erroneous in a complex world, and will lead to conflict over access to resources and distribution of their costs and benefits. Conflict is thus a *natural* part of the world that the design of institutions can mediate but never eradicate. The existence of conflict indicates that institutional design has thus far *failed* to account for and address the preferences of all interested parties, whether these are conceptualized as citizens, resource users, stakeholders, or other similar subject positions. This is a problem of faulty information exchange: individuals are unable to communicate their preferences to each other and to planners, which results in institutional arrangements that do not reflect preferences, inhibit information flow, and lead to conflict. Conflict compels researchers and practitioners to continue adaptively searching for satisficing outcomes that more appropriately align the complexity of individual preferences with delivery of services, including ecosystem services (Arrow *et al.* 1995; Folke *et al.* 2007).

Public choice theorists' understanding of institutional design thus renders conflict an object of rationalized management. This is a Simonian rationality of adaptively searching for satisficing outcomes, a search process that occurs through manipulating elements of the choice environment that can shape decision-making outcomes. In turn, the manipulation of these elements exercises an environmental form of power, which attempts to re/design higher-order rules that govern decision-making processes and collective behavior – and thus reshape the contours of the social itself.

Resilience theory absorbs this neo-liberal understanding of the social as an emergent effect of institutional design and constitutional choice. Elinor Ostrom (1999: 519–520) describes the process of designing environmental governance institutions in telling terms:

> Instead of assuming that designing rules that approach optimality, or even improve performance, is a relatively simple analytical task that can be undertaken by distant, objective analysts, we need to understand the policy design process as involving an effort to tinker with a large number of component parts. Those who tinker with any tools—including rules—are trying to find combinations that work together more effectively than other combinations.

Here, Ostrom is describing an indeterminate process of designing new institutional arrangements that combine competing interests and diverse forms of knowledge in new ways, and learning from – that is, tinkering with – these combinations to find provisional satisficing solutions. As we saw above in Richard Buchanan's (1992) design theory concept of *placement*, playing with the functional combinations of different components can generate new meaning, identity and function out of nothing other than the existing arrangement of things. Ostrom's designerly process of tinkering thus promises to generate new functional configurations that change how information flows across different system components – both social and ecological. Resilience scholars, taking their cue from Ostrom and other institutional economists, share this designerly sensibility. As we saw above, Per Olsson and colleagues describe social and ecological transformations as outcomes of functional recombinations: "transformability means defining and creating novel system configurations by introducing new components and ways of governing SESs [social and ecological systems]... Transformations fundamentally change the structures and processes that alternate feedback loops in SESs" (Olsson *et al.* 2004: unpag.). Carl Folke and colleagues similarly signal this designerly sensibility when they suggest that,

> transformations do not take place in a vacuum, but draw on resilience from multiple scales, making use of crises as windows of opportunity, and recombining sources of experience and knowledge to navigate social-ecological transitions from a regime in one stability landscape to another,
>
> (Folke *et al.* 2010: unpag.)

These understandings of transformation blend a designerly style of critique with ecological resilience theory: transformation occurs through the process of recombining existing adaptive capacities – "resilience from multiple scales" – and diverse forms of knowledge in order to create a new system configuration.

The terms *tinkering*, *learning*, *recombining* and *placement* all signal the same practice of designerly critique: generating novelty out of the existing arrangement of things through abstraction and functional synthesis. However, the residual methodological individualism that sneaks in through the backdoor of public choice theory leads resilience scholars to emphasize how institutional arrangements impact *information flow* in ways that affect *management and resource use decisions*. Just as public choice theorists sought to align delivery of public services with a complex array of individual preferences, so too do resilience scholars seek to align management and resource use decisions with complex ecosystems. For example, Folke and colleagues (2003: 370) suggest that "institutions organized in ways that parallel the structure of ecosystems are more likely to receive accurate and timely information about the state and dynamics of the systems and be able to respond in constructive ways." Indeed, provisioning high-quality information on dynamic system conditions is a paramount requirement of adaptive governance: Thomas Dietz, writing in an influential early study on adaptive governance with Elinor Ostrom and Paul Stern, suggests that, "environmental governance depends on good, trustworthy information about stocks, flows, and processes within the resource system being governed, as well as about the human–environment interactions affecting those systems" (Dietz *et al.* 2003: 1908). The problem is that aggregated information could "ignore or average out local information that is important in identifying future problems and developing solutions" (ibid.).

In a complex environment, all information offers potentially useful insights on system performance. The demand for this knowledge is practically and pragmatically unlimited – and thus governance is an unending problem of re/designing institutions to improve information flow and facilitate more effective decision-making (Chandler 2014a).

Rather than an expression of underlying power relations, conflict becomes one vector of information flow amongst others, a form of cybernetic feedback that conveys mismatches between institutional design and system performance. Conflict becomes useful as it becomes an indicator of opportunities to intervene in a system in order to "improve" (cf. Latour 2004) that system's functionality. As a mechanism for drawing attention to potentially hidden or unrecognized conflict, critical analyses can signal points of tension and misalignment that bottleneck information flow. Critique thus becomes an essential component of effective adaptive governance, for it demonstrates where power relations distort systemic functionality. For example, in their defense of resilience theory against charges that it cannot account for power and politics, Per Olsson and colleagues make an implicitly designerly argument that "the problem ... is not a lack of individual and societal innovative and transformative capacity, but rather how this capacity can be used to solve social and environmental problems and create the conditions for human welfare both today and in the future" (Olsson *et al.* 2014: unpag.). In their reading, the use of this capacity is constrained by barriers to transformation that are often social in nature, such as "culture and cognition and expressed through economic and social policies, land-use legislation, resource management practices, and other institutions and social practices'" (O'Brien 2012, in Olsson *et al.* 2014: unpag.). These social barriers impede information flow: they inhibit decision-making possibilities and lock systems into resilience traps, as above. The critical move, for resilience proponents, involves redesigning the social itself: transgressing social barriers through re-combining existing capacities in ways that make socio-ecological transformation possible. Conflict and critique thus become mechanisms for regulating information flow within a system: effective governance will recognize them as indicators of social barriers that impede information flow to managers and other decision-makers.

As such, resilience scholars have a longstanding interest in social and ecological difference. For example, different forms of knowledge on the social and natural worlds, such as indigenous worldviews or local peoples' intimate knowledge of their environments, can provide important correctives to abstract, technical knowledge that typically informs environmental management initiatives. But their interest is mediated through a functionalist sensibility that places implicit demands on difference to be amenable to functional synthesis: this different knowledge must provide some kind of pragmatic utility to managers, resource users, and other stakeholders involved in environmental governance processes. That is, difference must become useful. To examine how resilience attempts to make becoming-otherwise become useful, the next section explores how resilience proponents engage with radically alternative forms of knowledge on human–environment relations.

MAKING DIFFERENCE USEFUL: SOCIAL MEMORY, LEARNING, AND TRADITIONAL KNOWLEDGE

As we saw above, Kai Lee's pragmatic slant on adaptive management foregrounded the importance of experimental *learning* as a way to generate knowledge about complex social and environmental situations that exceed individual comprehension. Similarly, public choice theorists filtered this Deweyan insight through a liberal ontology of individual choice to develop an understanding of learning as the essential characteristic of human being. Indeed, for Vincent Ostrom, "human beings might be said to be capable of *learning how to learn*" (Ostrom 1997: 131). Here, learning involves an individual identifying, reflecting on, acting within and, if experience warrants, transforming constraints in the environment to advance individual goals. Human life is thus artifactual because humans are capable of learning.

On the one hand, Ostrom's argument might be construed as post-liberal: it immerses the individual within the world, and offers a vision of choice as a dynamic process driven by learning. Rather than locating choice in the individual will, it is now located at the interface or threshold between interior and exterior, subject and world. But on another level, Ostrom offers a vision of individual learning as a designerly process: to be human is to learn how to identify the points of disjuncture between an interior (mental image of (complex) environments) and exterior (material reality of these environments) and bring these in line with each other. If we focus on the designerly qualities of Ostrom's thought, rather than its formal post-liberalism, we can recognize how this vision of learning makes critique – designerly critique – a natural process, and an essential part of what it means to be human. Indeed, learning enables humans to take advantage of constraints – to introduce variety and adapt to a dynamic world (Ostrom 1980). In other words, change is possible *because* the environment constrains human cognition; humans grow, develop and change as they reflectively and creatively engage with, transform, and transgress these constraints. Of course, this vision of change still operates within the liberal problematic of finitude and an uncertain future. For Ostrom, learning introduces a degree of indeterminacy into human affairs: "this very creativity places a correlative burden upon every member of the species to face an uncertain future. All of life is an effort to take best advantage of changing opportunities and possibilities" (Ostrom 1976: 254). But like other liberal biopolitical techniques such as risk management calculations (Foucault 2008; Dillon 2008), institutional design enables humans to chart a course for action that will allow for continued growth and development in an uncertain and indeterminate world.

The convergence between ecological resilience theory and institutional economics helped reconfigure adaptive management techniques around this neo-liberal and designerly vision of learning. For example, Frances Westley (1995), an early proponent of institutional analysis within resilience scholarship, follows the footsteps of Dewey, Hayek, Simon, Ostrom and others to argue that, "at the grass-roots or 'local' level, individuals in organizations constantly respond to subtle changes in their environments, and ... these responses represent sources of innovation and learning." Thus, the most important information on system performance and change will often come not from leadership in an organization, but from those closest to the field – salesmen and women, in the case of businesses, or practitioners in the field, such as "biologists and wildlife managers" in the case of environmental management (ibid.: 403). One of the key challenges for institutional design is to structure organizations in a way that takes advantage of this "tacit knowledge" or "ongoing learning" that is "an almost instinctive propensity of human organisms." (Wallace 1961; in Westley 1995: 401–402).

Resilience scholars had long recognized the importance of basing environmental management decisions on multiple perspectives (Holling 1978). But the emphasis Lee and Westley, among others, placed on the need to *learn* from different perspectives allowed ecologists to re-evaluate the utility of non-Western forms of knowledge on environmental change. In his influential work, *Sacred Ecology*, Berkes (1999) draws on human ecology research into indigenous peoples' ways of knowing and interacting with their surroundings to define what he calls *traditional environmental knowledge*. With this term, Berkes (ibid.: 8) casts indigenous worldviews as a pragmatic and adaptive body of knowledge about human–environment relations, communicated through culture. Subsequently, resilience scholars identified the functional utility of indigenous cultural practices. For example, taboos often functioned to regulate access to and use of scarce but valuable resources (Berkes and Folke 1998). Berkes' solo and collaborative work thus enabled resilience proponents to apprehend indigenous knowledge and other non-Western forms of knowledge about specific ecosystems as rational abstractions: potential sources of information on system performance that could be functionally synthesized with Western scientific and management knowledge in order to improve sustainability. In the introduction to their influential text on resilience and institutional change, Berkes and Folke (1998: 14) signal this abstraction and synthetic potential when they argue that, "traditional systems … represent many millennia of human experience with environmental management, and provide a reservoir of active adaptations which may be of universal importance in designing for sustainability." Thus, even though many scientists may apprehend traditional knowledge as "a mélange of truth and inaccuracy" (Anderson 1996, in Berkes and Folke 1998: 14), local knowledge is *potentially* compatible with scientific knowledge: "local and traditional knowledge and management systems should be seen as *adaptive responses* in a place-based context and a rich source of lessons for social-ecological adaptations" (ibid.: 13, emphasis in original). The problem, as Westley identified in a different context, is to design institutions that facilitate synthesizing traditional and scientific knowledge.

Work on institutions thus opened a space for resilience scholars to incorporate alternative worldviews and indigenous knowledge systems into environmental management (Folke *et al.* 2007). But as critics have noted, this came at the cost of reducing entire cosmologies into abstract and functionally substitutable "bodies of knowledge" (Nadasdy 1999; Watts 2015). Thus, even as resilience approaches make a space for the inclusion of difference, read here as alternative ways of knowing and living the environment, they also impose an implicit demand on this difference: a demand that different ways of knowing, experiencing, and living the environment be amenable to functional synthesis (cf. Povinelli 2011a). This demand reduces radical, incommensurable difference to nothing other than a distinct set of relays and feedbacks between society and environment – relays and feedbacks that can be artificially designed-into Western societies and environments to improve information flow, build adaptive capacity, prepare systems for change, and increase sustainability. As Berkes and Folke (1998: 18, emphasis added) suggest, "societies that have survived resource scarcities ... [are those that] adapt to resource changes and learn to *interpret signals* from the resource stock through a dynamic social-ecological process, thus developing flexible institutions to deal with resource management." The italicized phrase is key: for resilience scholars, indigenous knowledge systems and Western science are potentially compatible because they are, in essence, nothing more than information-processing systems that receive and interpret information from their environments. Thus, indigenous knowledge simply represents a different way of *sensing* the environment. In turn, the process of synthesizing alternative and Western forms of knowledge is a process of learning how to sense complex systemic dynamics in new ways. For example, Westley (1995: 400) argues that successful adaptive management requires "sense-making processes" that can open management to new stimuli and facilitate new kinds of action in response to these stimuli. Folke and colleagues similarly stress that, "learning that helps develop adaptive expertise ... and processes of sense-making are essential features of governance of complex social-ecological systems, and these skills prepare mangers for uncertainty and surprise" (Folke *et al.* 2005: 448).

The persistence of indigenous peoples, even in the face of social and ecosystem disruptions, demonstrates that sustainability is possible, *if society can learn to access and interpret signals from the environment in the proper way*. The value of indigenous knowledge and other alternative forms of knowledge comes from its ability to demonstrate alternative sense-making practices that could potentially be useful for resilience proponents' sustainability-building efforts – but its utility hinges on a practice of designerly critique. To realize its value, resilience scholars must abstract indigenous knowledge from its context and synthesize it with established Western forms of knowledge and institutions in order to create new ways of sensing and engaging with the world. A variety of environmental techniques facilitate this synthesis. As we saw above in the case of Jamaican disaster resilience, participatory activities such as focus groups, transect walks and other educational activities abstract local knowledge on disasters and recode it in terms of Western science, producing new ways of knowing and experiencing the world and new forms of desires and capacities in the process. Similarly, in the realm of urban resilience planning, the recent development of decision theaters utilizes longstanding emergency management techniques such as table-top and live-action simulation exercises (see Collier and Lakoff 2008) to bring together a variety of stakeholders, including government leaders, community organizations, scientists, and other private sector and non-governmental partners to collaboratively explore different future urban development scenarios. Qian Ye and colleagues (2017: 186) describe how the pathbreaking Decision Theater developed at Arizona State University facilitates new forms of sense-making through synthesizing diverse forms of knowledge:

> Designed to help decision makers manage complex issues and to develop a shared understanding of issues based on each member's area of expertise quickly and confidently, the Decision Theater provides a platform for participants to create a shared mental model of their problem domain... [and] communicate within and across disciplines. It enables decision-makers to see things differently, develop consensus, and make knowledge-based decisions.

The Arizona State Decision Theater is quite literally a product of cybernetic environmental techniques. As Ye and colleagues (ibid.) describe it, the Theater features a room with a 270° screen that can display "panoramic computer graphics or 3D visualizations" of complex and dynamic urban social, ecological and technological systems. These visualizations allow participants to play with a system's component "parts" in order to simulate possible future outcomes that changes in social, infrastructural or environmental policy might produce. As with all simulations (Anderson and Adey 2011), these exercises are designed to produce certain affective atmospheres, such as the confidence to make decisions in a complex and uncertain milieu. But in this case, if effective, the designed affects emerge through a collaborative process: "participants create and modify their models through group discussions until simulated outcomes are judged sufficiently useful to take action" (Ye *et al.* 2017: 186).

Environmental techniques such as participatory programming and decisions theaters build on resilience proponents' wider efforts to capture and activate what they call *social memory*. This concept describes the "long-term communal understanding of the dynamics of environmental change and the transmission of pertinent experience" (Berkes *et al.* 2003: 20). For Berkes and colleagues, social memory

> "captures the experience of change and successful adaptations. Social memory is the arena in which captured experience with change and successful adaptation, embedded in a deeper level of values, *is actualized* through community debate and decision-making processes into appropriate strategies for dealing with ongoing change.
> (Ibid.: 20–21, emphasis added)

The italicized term is key: in this formulation, social memory refers to the latent and uneven traces past experiences of social and environmental change leave on individuals. These are experiences of change that might not register as properly scientific, and thus may not inform management practices – as we saw above in terms of indigenous knowledge, the Western scientific observer might see these experiences as "preposterous nonsense" and an unusable "mélange of truth and inaccuracy" (Berkes and Folke 1998: 14). However, as Folke and colleagues (2005: 453) suggest,

> a collective memory of experiences with resource and ecosystem management provides context for social responses and helps the social-ecological system prepare for change. If experience embedded in institutions and organizations provides a context for the modification of management policy and rules, people can act adaptively in the face of surprise.

Thus, "a key challenge for adaptive governance during periods of change seems to be the mobilization of social memory" (ibid.). Participatory activities, collaborative scenario planning, and adaptive institutional re/design all attempt, in different ways, to mobilize – actualize – social memory and make latent experiences of change and successful adaptation legible and usable for current adaptation efforts.

Here, there is a striking convergence with, and divergence from, the Foucauldian formulation of *subjugated knowledge*, or marginalized experiences and knowledge of change that has not been formalized as either totalizing "Historical" knowledge or scientific discourse. As Chapter 2 detailed, subjugated knowledge is a pillar of genealogical thought: one purpose of genealogical work is to unearth these subjugated knowledges in order to show how the existing order of things rests on nothing more nor less than ongoing social warfare (Foucault 2003; Philo 2007). Subjugated knowledges offer a window on the potential to become otherwise – a potential that inheres in the present as a latent force of actually existing difference waiting to be actualized. As such, the category of subjugated knowledge is a key component of critical theory, for it offers a window onto the power relations that structure the present *and* directs attention to points where the established order might be transgressed (Gordon 1986). The category of social memory zeroes in on this same transgressive force – the "captured experience with change and successful adaptation," as above, that has the potential to produce new ways of sensing and managing environmental change. But resilience scholars relate to this force through a designerly sensibility. Their question is not how subjugated knowledge/social memory shows the contingency of the present. Rather, their concern is how subjugated knowledge/social memory might be utilized to create new institutional forms, management procedures, and sense-making practices. Folke and colleagues (2003: 370) are forthright on this point: "there has to be a potential for combining and recombining social memory, adding and filtering external influences ... as well as transferring knowledge and experience in space and time ... to enhance overlapping functions and redundancy for resilience." This process of combining, re-combining, adding, filtering and transferring is an indeterminate synthetic process: there is no transcendental determinant to guide decision-making on this question. Instead, environmental techniques such as participatory exercises and collaborative scenarios provide a space where scientists, local peoples, state officials, and other interested "stakeholders" can collectively experiment with different syntheses in order to produce satisficing adaptation strategies.[8]

ENVIRONMENTAL CRITIQUE BEYOND CRITIQUE?

Thus, just as ecological resilience transvalues conflict into a stabilizing force, so too does it transvalue subjugated knowledge into a force that can enhance adaptability. Through the category of social memory, resilience scholars abstract subjugated knowledge and render it amenable to functional synthesis with other forms of knowledge, often in other contexts, in the name of enhancing adaptive capacity and designing sustainability into social and ecological systems. Many critical scholars zero in on these transvaluations as evidence that resilience creates depoliticizing effects. And there is much to this argument. For example, the actualization of social memory expresses what Brad Evans and Julian Reid (2014) call a "forgetting how to die": if death, in a philosophical sense, refers to the transgression of limits and becoming radically otherwise to oneself, then resilience proponents' efforts to fold alternative forms of knowledge and practices into existing systemic arrangements are meant to prevent precisely this kind of radical transformation from occurring (Grove 2017). Indeed, this topological sense of changing everything without changing anything lies at the heart of ecological resilience theory. Brian Walker and David Salt (2012: 3) describe this plasticity in terms of identity: "people, societies, ecosystems, and socio-ecological systems can all ... be subjected to disturbance and cope, without changing their 'identity' – without becoming something else." But as we have seen in this chapter, "being subjected to disturbance and coping ... without becoming something else" hinges on the exercise of environmental techniques that render the force of difference – the embodied potential to become otherwise – as abstractions that can be functionally re-combined with one another to produce topologically identical transformations. This process of interiorizing the outside – making becoming-otherwise become useful – is what Jeremy Walker and Melinda Cooper (2011: 157) identify as a tendency in resilience thinking to "metabolize all countervailing forces and inoculate itself against critique." And through this inoculation, resilience theory turns critique into a mechanism of control that, as Kai Lee's (1999: unpag.) above-referenced quote indicates, it experimentally exercises over "a mixed system in which humans play a large, sometimes dominant role."

But at the same time, resilience theory's underlying designerly aesthetics suggests that the form of critique resilience exercises is one that conventional critical theory is ill-suited to engage with. As we saw above, much critical thought is calibrated in relation to a will to truth. Where a will to truth attempts to generate objective and seemingly "natural" truths about social and ecological reality, critical analysis shows that these truths are always partial. Hence the common maneuver within critical studies of resilience to show how this new form of truth-telling about a complex world facilitates the ongoing neoliberalization of social and ecological relations. While this is an essential critical insight, by no means is it the whole *story*. As we have seen, resilience is organized around a will to design. The truths it produces about the world, and the forms of control it exercises, do not have the same totalizing effects of a will to truth (even though a will to truth persists in some examples of resilience theory that offer "strong" ontological claims about the theory's explanatory power – see especially Holling 2001, 2004). Instead, resilience recalibrates the study of human–environment relations around a will to design that strives to render social and ecological reality amenable to functional synthesis. It mobilizes a designerly style of critique that attempts to generate novelty through re/combining systemic elements in novel arrangements that can generate new capacities, new desires, and new ways of sensing and managing the world. To be sure, resilience techniques create a space to fold excluded forms of knowledge into the policymaking process, and thus allow decisions to reflect the needs, interests and worldviews of those commonly excluded from the policy making process (Folke *et al.* 2007). And there is something to say for this enlarged policymaking process: more inclusive, design-based policy making can address longstanding concerns from marginalized communities, if it remains truly inclusive (Collier *et al.* 2017). Resilience thus may be a site for new kinds of politics in an indeterminate world (Grove 2013b; Nelson 2014) – *because*, rather than in spite of, its underlying designerly aesthetics.

From a designerly point of view, resilience does not negate critique or politics as much as it strives to make critique and politics function differently. Rather than destabilizing a totalizing vision and showing the contingency of the present, critique after design involves making partial and contingent arrangements function in new ways. Resilience approaches thus *amplify* the need for critique and conflict, but its environmental techniques refract their essence – the potential to become otherwise – through the prism of categories such as social learning, institutional design, traditional ecological knowledge, and social memory. The result – to the extent that these techniques can be effective – is a world open to design: an environment apprehended as nested, hierarchically organized complex social and ecological systems that is *potentially* sustainable – if experimental adaptive management and adaptive governance techniques can succeed in rationally organizing information flows and enhancing systemic adaptability. After resilience, politics and critique do not strive to reinvent the world, but instead open design opportunities within the world.

Resilience theory thus mobilizes a form of critique that largely does not register within conventional critical environmental scholarship. Its playful re-combinations of abstracted social and ecological elements *can* have depoliticizing effects, and *can* reinforce rather than challenge the prevailing political ecological status quo. But to a large extent, much critical scholarship on resilience latches on to these effects *because that is what critique is calibrated to sense and analyze.*[9] There is thus an entire domain of designerly practice that flies under the radar of critical analysis. Designerly processes of abstraction and synthesis are sites where the virtual potentiality critique identifies become actualized in some ways rather than others. As the next chapter will suggest, these are sites where worlds are formed and re-formed out of nothing other than the present arrangement of things. While much critical research to date suggests that this often results in environmental management initiatives that stabilize rather than transgress the political ecological status quo, the designerly aesthetic suggests this stabilization is always partial and contingent, and could be designed in *other* ways.

Recognizing the designerly influence on resilience theory thus poses a challenge to critical environmental studies: it emphasizes the *ethical* dimensions of critique.[10] That is, critique increasingly becomes an unanswerable question of how to comport oneself towards other humans and non-humans in an indeterminate world. Resilience theory provides one answer to this question: it asserts that social and ecological relations should be reconfigured in ways that enhance adaptability and sustainability. Set in the context of global capitalist political ecological relations, this often results in resilience initiatives that depoliticize and depotentialize responses to environmental change. But it is only *one* possibility among others. The challenge for critical thought after resilience – indeed, critical thought after the Bauhaus – is to explore how scholars, activists, and even state officials and non-state agents might collaboratively create *other* possible futures. The next chapter outlines some possible vectors for future research on social and ecological change in an indeterminate world.

FURTHER READING

The most straightforward introduction to the concept of biopolitics remains Michel Foucault's Collège de France lectures (Foucault 2003, 2007a, 2008), where he elaborates on both the concept of biopolitics and the genealogical approach he utilized to develop the concept. In contrast to the dense writing in his polished texts, the lectures are considerably more dialogical, instructive and experimental. They remain the single most important resource for readers looking for an introduction to the concept. Excellent overviews of the political and philosophical debates surrounding biopolitics can also be found in Timothy Campbell's *Improper Life: Technology and Biopolitics from Heidegger to Agamben* and Thomas Lemke's *Biopolitics: An Advanced Introduction*. Mitchell Dean's *Critical and Effective Histories: Foucault's Method and Historical Sociology* situates Foucault's work within the wider trajectory of Western thought since the Enlightenment, and thus draws out the stakes of Foucault's particular slant on critique. Luc Boltanski's *On Critique* provides a detailed overview of the different valences of critique that circulate in both academic practice and everyday life.

NOTES

1 Absolutism attempted to organize and value all activities in relation to their usefulness to the absolutist state. To be sure, absolutism was itself a political response to Europe's internecine wars of the fifteenth, sixteenth and seventeenth centuries (Pasquino 1978; Foucault 2007a). But it made unity of the state a precondition for peace, and thus fully subsumed the private sphere under the sovereign's ultimate authority. In contrast, the Enlightenment sought to ground authority in rationality – a new configuration of public and private, state and science, sovereign and subjects (Koselleck 1988).

2 Behavioral geographers' engagements with Simon help illustrate the distinctive way resilience scholars in ecology picked up Simon's thought. In geography, scholars such as Robert Kates (1962) and Julian Wolpert (1964, 1965) drew on Simon's theories of bounded rationality and satisficing decision-making to critique prevailing theories of spatial behavior. Theories such as the gravity model of migration and distance–decay models were based on assumptions of a universally applicable economic rationality (Johnston 1979). Simon's thought thus provided behavioral geographers with a powerful framework to understand spatial patterns as results of boundedly rational, satisficing behavior. But they never engaged with the full implications of Simon's work. Ron Johnston's (1979: 117) description succinctly describes how geographers deployed Simonian thought:

> By studying behavioral processes in [specific] contexts, the aspiration was to increase geographers' understanding of how spatial patterns evolve, thereby complementing their existing ability to describe such patterns. Morphological laws and systems are insufficient of themselves for understanding; the amalgamation of concepts about decision-making taken from other social sciences with geography's spatial variable would allow development of processes theories that could account for the morphologies observed.

Simon's behavioral science provided behavioral geographers with a way to increase their understanding of how spatial patterns formed and changed –

and thus enabled them to fold a greater swath of social reality into a calculative will to truth. In contrast, as the previous chapter demonstrated, resilience proponents reconfigured the practice of ecological science around a will to design.

3 Gilles Deleuze (1995b) speaks to the effects of environmental power when he refers to the emergence of what he calls *control societies* in the years following World War II. For Deleuze, control societies do not operate on individuals – either the "individual" of disciplinary power or the "individual citizen" of liberal governmental rationalities. Instead, they operate on what he calls *dividuals*: abstracted characteristics that can be systemically configured in ways that enable certain kinds of circulation, growth and development and inhibit other forms of circulation. The paradigmatic example of control societies is perhaps regulation through code: scan a card or a fob (or, increasingly, a fingerprint, or a retina...) embedded with a specific code and gain entry to a building, room, files. Mechanisms of control, in this sense, modulate complex and emergent milieus: they allow some forms of emergence while inhibiting others from materializing. Control thus operates on the abstract, virtual *potentiality* that inheres within any system.

4 In this sense, the abstractions produced through designerly thinking are distinct from the abstractions produced through conscious thought. For the latter, abstraction involves a reductive process of slotting embodied experience into pre-formed, ideal categories. This is a process of moving from the specific to the general that allows the identification of transcendental essence. In contrast, the abstractions produced through designerly thinking are always oriented towards the relational capacities that inhere in an object. As cultural studies scholar McKenzie Wark (2004) writes in relation to hacking, to abstract is to "construct a plane upon which otherwise different and unrelated matters may be brought into many possible relations. To abstract is to express the virtuality of nature" (see also Chapter 7). Designerly abstraction here is the movement from contextualization to decontextualization, which frees existents to form new functional relations with others, thus giving them new identity, meaning and value. The abstractions produced through conscious thought thus centralize meaning around the categorical imperative; the abstractions produced through designerly thought thus disperse meaning across functional relations.

5 Dewey's arguments here on the relation between complexity and individual knowledge are a pragmatist precursor to both Simon's (1997) early formulations of bounded rationality and neoliberal economists' critiques of centralization (especially Hayek 1945). Importantly, Dewey (1938) – like Simon, Hayek, Holling and other resilience scholars – takes the complex nature of the world for granted. The problem is that complexity outstrips human capacities to know the nature of the world. Ontology is never questioned; instead, the problem is one of epistemology: how can the subject relate to a complex world? For Hayek, first-order cyberneticians and other theorists of "simple" complexity (Chandler 2014a), there is still a space outside the world where the subject can stand and judge the world in terms of universally valid transcendental determinants of meaning and value (for Hayek, this is the market, and the price mechanism in particular). For Holling, most resilience scholars,

second-order cyberneticians and other theorists of "general" complexity (ibid.), the subject is immersed within the world; its knowledge is immanent to the processes it seeks to understand (for example, this is Elinor Ostrom's (2007) oft-repeated claim that there are "no panaceas" to determining optimal institutional arrangements to govern common-pool resources. In contrast, as Chapter 7 will show, many radical scholars in geography, anthropology, cultural studies, international relations and design studies situate complexity as one particular ontological formation that emerges as scholars grapple with indeterminacy in any number of fields. Recognizing complexity's contingency suggests that ontology itself is malleable and thus political. The affirmation of ontopolitics is one of the distinguishing characteristics between radical scholarship on world-making in the Anthropocene and resilience theory, for the latter denies any possibility of remaking a complex world (see Chapter 7; Evans and Reid 2014).

6 James Buchanan received the Nobel Prize for Economics in 1986, served as president of the Mont Pelerin Society, a cradle of neoliberal economic thought, from 1984–1986, and was the inaugural president of the Public Choice Society in 1964. Vincent Ostrom served as president of the Public Choice Society from 1967–1969.

7 I use the hyphenated *neo-liberal* to distinguish the more unconventional strands of public choice theory from conventional neoliberal economic thought typically associated with Hayek, Friedman and Becker. If the latter emphasize market solutions to problems of complexity, the former emphasize how there is no single determinant of value that is universally applicable for all choice situations. As Stephen Collier shows in relation to both Buchanan (Collier 2011) and Vincent Ostrom (Collier 2017), what makes these writers neo-liberal is their effort to critically recalibrate key tenets of liberal political philosophy in relation to novel experiences of complexity and indeterminacy.

8 Of course, there has been a growing interest in experimental techniques to develop urban resilience, which has been driven in large part by urban ecologists (Grimm *et al.* 2000; Pickett *et al.* 2004; Grimm *et al.* 2008; Ernston *et al.* 2010). As a number of critical human geographers have shown, this experimental ethos has filtered into other arenas of urban social, economic and environmental governance (Evans 2011; Bulkeley and Castán Broto 2013; Karvonen and van Heur 2014; Evans *et al.* 2016). On the surface, there appear to be important similarities between this work and more general resilience theory, but the links between a will to design and urban experiments need to be more fully fleshed out.

9 This is a point Collier (2011, 2017) emphasizes in relation to critical scholarship on neoliberalism. While neoliberal political economists engage in their own critiques of centralized economic planning, critical scholars often pass over their critical attitude and strategies because they are not in an immediately recognizable form. Collier thus argues that critical scholars need to broaden their understanding of critique in order to understand contemporary configurations between knowledge and power, and the role critique plays in cementing as well as challenging the political economic status quo.

10 This ethical dimension is something Foucault explores in his focus on the "government of the self and others" in his later work. Starting in his 1979–1980 lecture series, Foucault drops the threads his biopolitical analysis of neoliberalism opened, and begins exploring "techniques of the self" in ancient Greece. Here, he focuses on *ethical substance*, or that within the human which endows it with the potential to become otherwise (see also Povinelli 2011a).

7

UN-WORLDING RESILIENCE

INTRODUCTION

Writing on practices of living with crisis, political philosopher Achille Mbembe and anthropologist Janet Roitman (1995) offer a stark account of everyday life in early 1990s Cameroon. During this time, the country was undergoing devastating disinvestment. The promise of unlimited, capital-driven growth that fueled speculative building during the late 1980s gave way instead to a ruined social and physical landscape. Construction projects sat half finished, and routine events such as paychecks or office hours would occur at irregular intervals, if at all. Here, crisis is not a single moment of rupture, but rather a gradual, durative routinization of improvisation. Despite their immersion within enervating environments of neoliberal abandonment, people still strove to construct dignified, meaningful lives within *and through* this ruination. As Mbembe and Roitman detail, this required a variety of new capacities for living, such as a willingness to use whatever structures and apparatuses were at hand to advance their immediate, pressing goals. This could mean treating employment in a state office as a means for personal gain by charging under-the-table fees for services (especially important since there was no guarantee a formal paycheck would materialize at the end of the month). Or it could involve marking unwitting Western tourists as targets for various hustles. Whatever it might entail, surviving within the crisis required a certain kind of ethos – an attunement to the immediate, pragmatic functional utility of everything and everybody.

At first glance, this might appear to have little to do with resilience. But in a particularly insightful passage, Mbembe and Roitman suggest that,

> acting efficaciously requires that one carefully cultivate an extraordinary capacity to be simultaneously inside and outside, for and against, and to constantly introduce changes in the reading and usage of things, playing, in this way with the structures and apparatuses, capturing them where possible and eluding them where necessary, and in any event, amputating them and almost always emptying them of their formal and designated functions so one can better restore them with those that correspond best to desired goals and expected gains. (ibid.: 340)

Put in terms such as "playing," "constantly introduc[ing] changes in the reading and uses of things" and being "simultaneously inside and outside," these strategies for "acting efficaciously" bear a number of topological similarities to ecological resilience theory's designerly sensibilities. Both share an engagement with the world from positions of limited and partial knowledge. For ecologists, this is the lack of predictive certainty; for many Cameroonians, the immediate need to survive in the precarious life-world of neoliberal abandonment. In both cases, there is a lack of transcendental guarantees: the teleological promise of development and growth no longer holds sway in Cameroon; while ecologists cannot manage complex systems through techniques of optimization. Thus, resilience scholars and many Cameroonians grapple with the same kind of problem: how to *design* life in conditions of indeterminacy, or how to act in ways that align interiors (management goals and individual survival, respectively) with exteriors (complex social and ecological systems and the immediate social and physical surroundings). But at the same time, there are important differences: Mbembe and Roitman's account of many Cameroonians' inventive and creative functional re-purposing of everyday life does not hinge on cybernetic understanding of life as information exchange. Nor does it rely on a Simonian ontology of hierarchically organized complexity and bounded rationality.

Mbembe and Roitman's description thus lays bare a key question that has implicitly underpinned this book – what is at stake in the concept of resilience? Given that creative re-purposing may be practiced and grasped in a variety of ways that do not necessarily share the same vision of complexity as resilience theory, what is gained and what is lost by apprehending them through the concept of resilience? Or, stated in more explicitly ethico-political terms, what possibilities for living with indeterminacy does resilience open, and what possibilities does it foreclose?

This is a key challenge that resilience poses for critical research in geography and allied disciplines: given the ubiquity of resilience, what are the possibilities for critical thought beyond a resilience paradigm? How might we practice critique in a way that works against resilience theory's tendency to make becoming-otherwise become useful? In this chapter, I will suggest that critical geographic thought after resilience hinges on the form of topological thought that critique mobilizes. As Chapter 3 discussed, topological thought is attuned to processes of change and becoming; it is situated at the threshold of inside and outside, the real and the virtual, or the potential for things to become other than they are. If resilience interiorizes the force of the outside, then the critical challenge is to mobilize forms of thought that turn critique back to the outside, and that open thought up again to potential for novelty.

The chapter begins with a discussion on topological thought in geography and cognate fields. This shows how topological thinking can open a back door onto structural analysis that impedes our ability to recognize social and ecological difference as difference per se, even as it orients thought towards relationality and dynamic processes of change and becoming. With these stakes outlined, I then show how resilience approaches deploy topological thought to render socio-ecological difference knowable through the series Sustainability–Institutional Design–Learning. This means that resilience overcodes indeterminacy as a problem of uncertainty that threatens sustainability, which can be managed through the process of institutional design and feedback-driven adaptive governance. Resilience thus reduces the experience of indeterminacy to something that can potentially be managed through environmental techniques. This is the point where resilience interiorizes the outside. To work against this interiorization, I signal three alternative areas of work that turn thought back to the outside: work in geography and anthropology on the geopolitics of the Anthropocene; work in cultural studies and international relations on hacking and hacktivism; and work in radical design studies on transition design. While each of these areas can easily feed into a resilience agenda, it also gives us a sense of possible research directions that go beyond resilience. The series Anthropocene–Hacking–Transition Design thus provides one possible way for scholars in geography and cognate fields to engage with indeterminacy that is irreducible to the regulatory series of Sustainability–Institutional Design–Learning.

THE STAKES OF TOPOLOGICAL THOUGHT

Elizabeth Grosz directs our attention to the stakes of topological thought when she contrasts the thought of Jacques Derrida and Gilles Deleuze in the following terms:

> Where Derrida could be described as the philosopher who insists on *bringing the outside, the expelled, repressed, or excluded, into the inside* by showing the constitutive trace it must leave on that which must expel it... Deleuze could be understood as the philosopher who *evacuates the inside... forcing it to confront its outside*, evacuating it and thereby unloosing its systematicity or organization, its usual or habitual functioning, allowing a part, function or feature to spin off or mutate into a new organization of system, to endlessly deflect, become, make.
> (Grosz 2001: 71–72, emphasis added)

The italicized passages are key: Both Derrida and Deleuze stake their philosophies on reconceptualizing the relation between interiority and exteriority. But they take different tacks to the problem. Deleuze's thought of the virtual continually opens an interior to its outside: it can always become other than it is because the outside constantly provokes reconfigurations. In contrast, Derrida's *différance* interiorizes the outside: the outside is always already part of the inside. The outside is not generative, as in Deleuze, but rather de(con)structive: *différance* unbundles any identity, leaving it in ruin.

Grosz thus drives home the point that while topological thinking blurs inside and outside, *how* this blurring occurs is an open-ended question, and can often work in ways that limit the ability of critical thought to engage with difference as difference – that is, difference whose content is not pre-determined (see Povinelli 2011b). Given the growing interest in topological modes of thought in geography and allied fields (see Box 7.1), her argument deserves careful consideration. As Lauren Martin and Anna Secor (2013) demonstrate, topological thought can inadvertently introduce a back door into structuralist analysis that accounts for socio-spatial change and continuity in terms of an overarching, transcendent determinant. This is a point Sallie Marston and colleagues (2005) echo in their Deleuzo-Guattarian critique of relational approaches to scale. In their reading, scalar thinking in geography operates through a more or less implicit hierarchy. Here, hierarchy is not a matter of prioritizing one side of a binary over another; instead, it signals a structuralist mode of thought that attempts to account for phenomena through reference to an overcoding signifier. This often takes the form of a mode of explanation that examines how the local expresses and makes manifest more fundamental (and determining) global processes. In this case, while topological thinking may blur the boundaries between local and global (Amin 2002; Allen 2011) and enable geographers to envision place as an emergent confluence of multiple, multi-scalar trajectories (Massey 2005), it can also leave in place a form of analysis that directs the critical gaze upwards towards a privileged determining force – whether this is capitalism, globalization, sovereignty, discourse, or some other transcendent explanatory variable. As a result, even relational and topological accounts of place and space can still smuggle in pre-packaged structuralist explanatory frameworks.

Box 7.1 TOPOLOGICAL THOUGHT IN GEOGRAPHY

Many human geographers have more or less explicitly drawn on topological thinking since at least the mid-1990s "relational turn." Topology provides geographers with a spatial metaphor that moves beyond scalar and territorial imaginaries that assert an ontological division between space and place. For example, Ash Amin's (2002) topological understanding of globalization emphasizes how new information and communication technologies produce new senses of connectedness (and disconnectedness) as they fold together (or tear asunder) distant and proximate people and resources. In this and other accounts of globalization (Massey 2005), economic geography (Yeung 2005), neoliberal development (Peck and Tickell 2002), power (Allen 2003) and nature–society relations (Whatmore 2002) – to name a few – topology enables geographers to rethink binaries such as near/far, space/place, local/global, nature/culture and inside/outside not in absolute or relative terms, but rather as effects of ontologically prior social, political, economic, and affective relations. In these readings, space and place are immanent to social processes. That is, rather than serving as a pre-existing backdrop to social activities, they are produced as these processes play themselves out.

Topological thought has thus enabled geographers to progressively develop relational and process-based understandings of space and spatiality that challenge structuralist characterizations of space as a static domain of fixity and being and time as the dynamic domain of change and becoming (Massey 2005; see also Soja 2011). This is especially clear in Doreen Massey's (2005) account of space as the site of contemporaneous multiplicity, the site of juxtaposed spatio-temporal trajectories that prevent any essentialized, objective understanding of space and place and imbue space with the potential to become other than it is. Here, topology signals a heterotopic understanding of space, a sense of space as constantly in motion, constantly undergoing (de)formation (Lash 2012). Rather than assuming topographic division, quantitative distance and essence, topology focuses on contingent practices of in-folding, transduction, and, more generally, uneven processes of becoming. Thus, it is often taken for granted as an inherently critical and radical form of thought – even as recent appraisals caution against such assumptions.

This line of critique highlights an important yet somewhat underappreciated stake in topological thought. Marston and colleagues (2005) draw our attention to the way that topology can be mobilized to create and bound new objects of truth and new forms of certainty in ways that continue to erase lived, embodied and situated difference. The problem here is not the production of objects of truth per se, but rather that the way these objects of truth are produced can continue to blinker critical analysis – and thus our ability to imagine new political possibilities. At issue is *how* geographers deploy topological thought. Topology can enable geographers to break free from structural and territorialized determinants of subjectivity and identity, and thus engage with difference as difference. But it can also enable geographers to rethink how structure operates in more fluid, dynamic, and embedded ways, and thus continue to submit difference to the demands of identity.

The question of how geographers deploy topology is a question of how thought relates to the outside – that is, to the potential to become otherwise. A defining characteristic of structuralist thought is the creation and deployment of a topological space that enfolds everything within a pre-determined series – or structure – of roles and identities (Deleuze 2004). This is a topological space that continually interiorizes the outside, that renders entities visible, knowable and nameable by virtue of their relative position within a structure. This structure may itself be mutable, but what matters is its continued production of order out of disorder: its ability to overcode phenomena in a way that continually gives everything meaning, value and significance in terms of a pre-determined, transcendent semiotic hierarchy.

These debates over topological thought in geography and related fields should thus add another layer of caution against any quick assumptions that resilience theory's topological approach to environmental science and management is necessarily radical. As this book has shown, resilience names a particular *style of intervening* in an indeterminate world. It gives a certain kind of sense to experiences of spatial interconnection and temporal emergence that makes these experiences legible as expressions of complexity, and renders them amenable to governance through environmental techniques of adaptive management and adaptive governance. The next section explores how these interventions structure the possibilities for meaning in a way that interiorizes the force of the outside.

THE STRUCTURE OF RESILIENCE

Gilles Deleuze's transcendental empiricism helps us understand how topological thought constructs a meaningful world. For Deleuze (1990), sense (and non-sense) emerges through a process of serialization. This refers to the way we inscribe immanent, contextually specific phenomena into mental (or ideal) categories of meaning that approximate, but do not fully capture, some attributes of those phenomena (see Chapter 6 note 4). These abstract categories enable us to slot embodied experiences into meaningful "worlds" that make intuitive sense. Habits, memory, categories and other mental faculties allow certain phenomena to blur into the background of the normal, expected rhythms of everyday life, and heighten our sensitivity to other phenomena that stand out as remarkable, abnormal, or otherwise requiring our attention. But as Deleuze emphasizes, these categories also fail to capture the totality of immanent, embodied experience. There is always more to the world than we can consciously apprehend. Something, someone, can always potentially be more than it is, more than it has been. New meanings, values and identities autonomously emerge as people engage with one another and their surroundings. Serialization is thus a process of bracketing embodied experience in certain ways rather than others. It is a reductive process that creates order out of disorder and transcendence out of immanence.

As one style of intervening in an indeterminate world, resilience attempts to structure experiences of indeterminacy through categories such as complexity, multiple equilibrium systems, panarchy, the adaptive cycle, adaptive management, adaptive capacity, social learning, self-organization, social memory, traditional environmental knowledge, adaptive governance, transitions and transformation, to name a few. These concepts form signifying chains that hierarchically straitjacket potential meaning and value around three attractors: sustainability, institutional design, and learning. This is the overcoding structure of resilience: a regulatory series, Sustainability–Institutional Design–Learning, that apprehends social and ecological phenomena in terms of the problem they pose to future growth and development prospects (sustainability), their actual and potential functional utility for addressing this threat (institutional design), and the mechanisms for solving this problem (learning).

First, the cybernetic vision of *sustainability* encodes social and environmental phenomena in terms of how well they enhance or diminish adaptability. As we saw in Chapters 3 and 4, resilience proponents' topological sensibility enabled them to re-conceptualize sustainability in terms of linked social and ecological systems that stretch from the local to the global. In their hands, sustainability became a question of systemic resilience: how can a society maintain the conditions for future growth while growing and changing in the present? Rather than homogeneity, efficiency and optimization, sustainability requires, *inter alia*, flexibility, diversity, redundancy, responsiveness and adaptability. Resilience approaches thus heighten sensitivity to those attributes of social and ecological phenomena that enhance or diminish these desirable qualities. For example, certain forms of traditional ecological knowledge may provide scientists and practitioners with useful ways of sensing environmental change and thus enhance information flow between managers and complex ecosystems. In contrast, other forms of knowledge, such as animist religion or political economic explanations of vulnerability, may distort information from the environment through mysticism or detached structuralist explanation, respectively. Similarly, command-and-control approaches increase homogeneity, efficiency and optimization, which actively undermine systemic resilience. Resilience scholars label social and ecological systems organized around these forms of knowledge maladaptive, and target them for strategic programs of transformation.

Second, if sustainability names the problem-space of resilience and identifies objects of governmental intervention, then *institutional design* names the solution to these problems. Chapter 4 detailed how resilience proponents visualize institutions as the mediating link between complex social and ecological systems. Institutions perform a cybernetic function that is intimately connected to sustainability: they facilitate or inhibit information exchange across complex social and ecological systems, and thus enhance or erode adaptive capacity. For example, dominant cultural institutions such as self-interested consumerism and anthropocentrism, doctrinaire beliefs in efficiency, profit maximization and utility maximization, and management practices based on positivist science and command-and-control strategies reinforce a narrow utilitarian view of the environment, distort information from the environment, and thus lead to unsustainable practices. As we saw in Chapter 5, this cybernetic vision expresses an underlying modernist design aesthetic that grasps what exists as nothing more than signs denoting functionality. The Bauhaus school consolidated this aesthetic as it blurred art and life, prioritized function over form, and re-conceptualized ontology in terms of abstract functionality. As this design aesthetic seeped into new domains of scientific and managerial practice during the middle decades of the twentieth century, it facilitated the cybernetic reduction of life to information. Importantly, it also rendered social and ecological phenomena governable through environmental techniques.

We can see this reduction of life to information, and its biopolitical effects, in the subtle ways resilience programming attempts to transform how individuals can relate to one another and their wider environment. For example, institutional change often involves resilience initiatives designed to create resilient "cultures of safety" through transforming everyday norms that mediate human–environment relations (Chandler 2014a). Techniques of security and environmental power such as participatory education programs or insurance attempt to work on these norms and reshape how individuals and communities relate to both their surroundings and an uncertain future (Grove 2010, 2012, 2014a). For example, insurance encourages purchasers to see themselves as individuals in possession of capital whose future returns are at risk from hazard exposure. If the purchase of insurance cannot prevent a loss from happening, it can guarantee compensation when the loss does occur, which gives individuals confidence to act in an otherwise risk-filled world (Ewald 1991; O'Malley 2004). Similarly, educational programming and collaborative activities can encourage individuals to visualize themselves as part of larger systems. This could involve participatory activities that encourage individuals in a rural region to recognize themselves as part of a vulnerable community exposed to hazards (Agrawal 2005; Grove 2013a). Or it could involve collaborative planning initiatives that encourage representatives from different government agencies to recognize their mutual interdependencies and identify areas where their agencies could work together on common complex problems (Rogers 2012; Zebrowski 2016; Ye *et al.* 2017). In all cases, resilience initiatives attempt to transform the informal norms that implicitly shape how members of society interact with each other and the environment – and thus determine their functional capacities and identities.

The problem of sustainability thus boils down to a problem of designing institutions that structure information flow between complex systems and their environment. This brings us to the third over-coding attractor, *learning*. As we saw in Chapters 5 and 6, cybernetic control operates through feedback that measures the divergence between actual and desired system states. Resilience scholars blended this cybernetic mechanism with pragmatic philosophy's emphasis on experiential learning to inform their approach to adaptive management and adaptive governance. For both Herbert Simon and resilience proponents, adaptation is a rational, stepwise process of learning about complex environments from a position of necessarily limited and partial knowledge. In a hierarchically organized complex world, learning becomes the mechanism for surviving and thriving, and feedback is the mechanism that facilitates learning.

For resilience scholars, learning thus signals an inductive, stepwise and – as Chapter 5 showed – *rational* process of engaging social and ecological complexity. However, in contrast to positivist science and command-and-control management, the truths learning generates do not feed into strategies of predictive control. Instead, the partial and limited truths cybernetic learning generates become efficacious only as the institutional design process pragmatically synthesizes this truth with other, equally partial truths. As an ecological mode of designerly thought and practice, resilience approaches mobilize environmental techniques to reduce what exists to rational abstractions amenable to functional synthesis. And its success hinges on cybernetic feedback that facilitates information exchange, learning, and ongoing, adaptive adjustment.

This cybernetic vision of learning transforms the meaning and value of expressions of difference. Difference is essential to resilience, for it provides the motor that drives forward adaptive and transformative change: we learn because we are exposed to social and ecological phenomena we did not expect to encounter. However, environmental techniques place an ethical demand on social and ecological difference. Difference must make itself legible in certain ways rather than others; specifically, *difference must present some kind of functional utility to others.* Resilience approaches judge different forms of knowledge, such as local knowledge or indigenous knowledge, on the basis of the utility they provide to scholars seeking creative, integrative solutions to intractable problems of complexity. They judge different ways of engaging with environments in terms of the abstract "lessons" these can provide scientists, managers and other resource users in other contexts. In turn, they deem different forms of knowledge and practices that do not offer some kind of pragmatic utility maladaptive, mark them as threats to sustainability and target them for corrective improvement through participatory training, collaborative planning, and other resilience-building techniques. Similarly, as Chapter 6 demonstrated, the influence of pragmatism and public choice theory on resilience theory transformed conflict into a form of feedback. Rather than an expression of irreducible social and ecological difference, conflict indicates a possibility for learning: it signals where one individual's or group's interests diverge from the system's delivery of services. Conflict thus indicates a design opportunity, a gap between interior and exterior that can be addressed through re/designing institutions that mediate between preferences, interests and services provisioned.

This demand for synthesis thus flattens out difference in the name of integrative, pragmatic sustainability solutions. In this sense, resilience-based approaches interiorize the force of the outside: they delimit the possible meanings and values of different ways of knowing and engaging with the environment, and work on this difference to incorporate it into the overarching biopolitical problem of building sustainability and resilience.

DIAGRAMMING THE WORLD OF RESILIENCE

Taken together, the series Sustainability–Institutional Design–Learning structures the possibilities for engaging with indeterminacy. It transforms an immeasurable and unknowable potential to become otherwise into a determined series of social and ecological possibilities. In this sense, resilience approaches can be seen as enacting a kind of "worlding:" drawing certain existents into mutually constitutive (and in this case, functional) relations with each other; forcing others apart, and thus delimiting the possible meanings, values, functions and identities these existents might assume.

One illustrative way to think the coordinates of this world is as a two-dimensional vector field (see Figure 7.1). As we saw above, meaning is dispersed across two axes: a horizontal problem-solution axis (or equally, maladaptation-adaptation); and a vertical concrete-abstract axis. Value is dispersed from the lower-left corner to the upper-right corner. The upper-right quadrant, representing phenomena that are functionally abstract and adaptive, are the resilience ideal: these are conditions that contribute to sustainability and are amenable to functional syntheses that can create new sustainability solutions. The lower-left corner is most problematic: these are conditions that are contextually dependent maladaptations, which will continually erode sustainability. We could array different

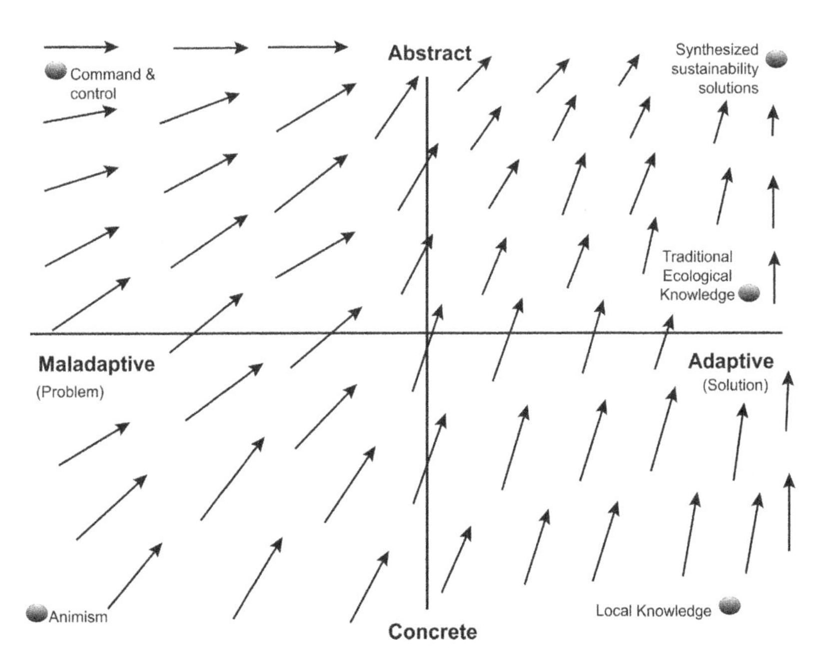

Figure 7.1 The world of resilience

phenomena throughout this field. For example, animist explanations of social and ecological insecurity might be located near the lower-left corner: they are context-dependent and maladaptive distortions of ecological signals, according to resilience accounts. Political economic explanations of vulnerability might be located along the left-hand side: they are somewhat context dependent, because they refer to specific political economic situations, but their structuralist explanations have elements of abstraction. They also distort information on ecological and social signals, which makes them maladaptive as well. Command-and-control management strategies might be located in the upper-left corner. They rely on optimization calculations, so they are abstract, but they distort information flow, so they are maladaptive and undermine sustainability. In the lower-right region we might find certain forms of local knowledge and indigenous knowledge: they are contextually specific, but also capable of capturing signals from complex ecosystems, which makes them adaptive.

Resilience techniques associated with adaptive management and adaptive governance provide the direction of force in our vector field. The movement is generally one from maladaptive problem and concreteness to adaptive solution and functional abstraction. For example, participatory techniques such as transect walks or focus groups can capture local knowledge and fold it into resilience plans, moving local knowledge from concrete and adaptive to abstract and adaptive. Educational techniques can attempt to replace animist or political economic explanations of vulnerability and insecurity with those based on hazards or climate science, moving these explanations from maladaptive and more or less concrete to adaptive and abstract. And adaptive management techniques designed to increase collaboration between scientists and managers and fold in feedback can transform abstract and maladaptive command-and-control management practices (those located in the upper-left quadrant of Fig. 7.1) into adaptive and abstract adaptive management practices (those located in the upper-right quadrant of Fig. 7.1).

If we wanted to more faithfully illustrate this world of resilience, we could add a third dimension, a fitness landscape of institutional change. Here, the force of the vectors would be inflected by the amount of institutional change required to transform the maladaptive and concrete into adaptive and abstract. For example, we might expect the barriers to adaptability – or the amount of change to undergo – to be highest in the upper left and lower left corners: in the examples we have been using, these represent people and communities whose personal and/or professional identity is constitutively tied to the formal and informal institutions of command-and-control management and animst worldviews, respectively.

Whether two- or three-dimensional, this vector field illustrates the underlying structure of resilience theory, which attempts to ascribe meaning and value to the totality of social and ecological phenomena (Holling 2001, 2004) and do so in a way that immediately opens them up to biopolitical regulation: either they are already abstract adaptive solutions, and thus capable of feeding into future systemic resilience, or they are concrete and/or maladaptive, which subjects them to certain forms of intervention (or directional vectors) designed to make them more adaptive and given to functional synthesis. And the animating force in this world, the meta-vector that gives all other vectors their force and direction, is a will to design that strives to render what exists as rational abstractions amenable to functional synthesis – in this case, in order to produce pragmatic sustainability solutions out of the more or less unsustainable order of things.

In more philosophical terminology, this movement from maladaptive-concrete to adaptive-abstract is the movement of interiorization that makes becoming-otherwise become useful. Expressions of social and ecological difference are not confined to a specific region of this vector field, but wherever they percolate to the surface of perceptibility, they are immediately swept up in the force fields that flow from maladaptive-concrete to adaptive-abstract. Through the series Sustainability–Institutional Design–Learning, resilience approaches make difference legible in ways that immediately render it governable and thus potentially usable within a cybernetic economy of designing sustainability solutions. This is how resilience interiorizes the force of the outside: environmental techniques turn expressions of social and ecological difference that point to *other* possible worlds into drivers of creative adaptations that open onto more sustainable systemic transformations. Resilience thus amplifies the need for politics and critique in a complex world, but it makes politics and critique function differently: rather than destabilizing the present, politics and critique become the means for securing an adaptive systemic order in an already destabilized and insecure world.

The possibilities for critique in a complex world thus hinge on transgressing this Sustainability–Institutional Design–Learning series that structures the world of resilience and makes becoming-otherwise become useful. In the remainder of this chapter, I want to outline one alternative series that might orient topological thought to the outside. This series is Anthropocene–Hacking–Transitions Design.

WORLD-MAKING IN THE ANTHROPOCENE

As Chapter 2 detailed, Brad Evans and Julian Reid (2014) locate an Anthropocene imaginary at the heart of resilience as a form of "non-death." The vision of a vengeful and fascistic Earth that destroys the life that depends upon it compels people to turn in fear from social and ecological potential to become otherwise. Following Holling's (1986) vision of the adaptive cycle, resilience scholars apprehend potentiality as the possibility for systemic breakdown – the Ω-phase of creative destruction. This is the point where sustainability becomes a problem: how can we ensure continued growth and development given the indeterminate potential for catastrophic systemic breakdown? As we saw above, resilience theory offers one response to this problem – and its roots in cybernetics and modernist design aesthetics, and its resonance with neoliberal security rationalities make it particularly compelling to many individuals and state officials.

However, geographers' engagements with the Anthropocene suggest a more complex suite of political possibilities. While not downplaying how a host of sovereign, biopolitical, environmental and security techniques are emerging to manage the qualitatively distinct insecurities and potentialities of the Anthropocene (Dalby 2013; Clark 2014; Grove and Chandler 2017), this work also recognizes how an Anthropocene imaginary opens onto new ethical and political problems. For Nigel Clark (2010), the Earth's indifference to human life reveals the radical contingency of life itself (see Box 7.2), and the non-reciprocal debt we owe to both human and non-human forebears who have bequeathed a liveable world to us. By this, Clark means that a liveable, inhabitable planet has been *artificially fashioned* over millennia through a host of non-human organic and inorganic processes, and the efforts of early humans to devise ways of sensing and living with these processes. Their actions have molded a world that we are able to apprehend as given to human use.

Box 7.2 THE RETURN OF KANT'S SUBLIME

On the morning of 1 November 1755, an earthquake now estimated at approximately 8.5 on the Richter Scale struck off the coast of Lisbon, Portugal. Pious residents who had packed churches for All Saints' Day worship were trapped inside as the ground began to shake and buildings began to collapse. Fires broke out across the city as candles lit for worship in churches and homes set buildings and rubble ablaze. Many residents who survived the shaking emerged in the streets to survey the damage, only to be swallowed by a tsunami that made landfall 40 minutes later. In its wake, the earthquake and tsunami left behind roughly 10,000 dead and 40–50,000 wounded – nearly half of the city's population – and a scarred landscape (Chester 2001). But the event had world-historical impact as well. Its senseless and arbitrary death and destruction broke down the coordinates of theodicy that had structured the classical European worldview. The horror of Lisbon prompted thinkers at the time to grapple with the reality of random and indiscrete violence that resulted from chance rather than divine retribution. In particular, Lisbon confronted thought with what Immanuel Kant called the *sublime*: events of such magnitude that they cannot be grasped or comprehended by individual consciousness (Clark 2010). The sublime signals what exists beyond thought – a world of unimaginable, indeterminate forces that create the most breath-taking beauty as well as the most horrific suffering. The sublime is thus the exterior that bounds human consciousness. It names an immeasurable potential that follows no laws human politics, science or theology can create. Of course, Kant's solution when faced with the sublime was to retreat into the shelters of self-conscious certainty. Kant thus attempted to replace the indeterminacy of the world with the transcendence of the categorical imperative. This withdrawal from the world helped constitute the modern subject of sovereign will and interests, hermetically sealed off from the wider world.

But if this barrier between subject and world ever truly held in European culture, it began cracking in the early 1900s. The crisis of technology and rationality World War I provoked, industrial society's increasing dependence on large-scale, complex technological infrastructure systems, the creation of the atom bomb, recurring global economic recessions and eventually the collapse of the Cold War replaced the certainty and stability of borders and predictability with a growing sense of pervasive insecurity. Only instead of following Kant's retreat, the aesthetic regime of modernist design and the complexity sciences that grew out of it immerse life within this indeterminate world, providing a variety of artistic and scientific techniques for orienting the subject in a world beyond its comprehension. As Chapter 5 detailed, these insecurities sparked critical reflections on art and life (the Bauhaus), rationality (Simon), and command-and-control management styles (ecology) that gave rise to contemporary forms of resilience.

However, the more recent naming of the *Anthropocene* has intensified this indeterminacy. The term signals a new geological epoch in which human activity has become a driving force of planetary change. First popularized by the Nobel Prize-winning atmospheric chemist Paul Crutzen (2002), the concept has generated heated debate among physical scientists and social scientists alike (Malm and Hornberg 2014). Despite these arguments, for many theorists, the simple act of naming humans as a geological force already has profound consequences. As Nigel Clark (2010) patiently details, the Anthropocene imaginary destabilizes the ground of the modern subject. The term *ground* works on two levels. First, in philosophical terms, it refers to the clearing where "what exists" reveals itself as such, or becomes available for conscious reflection. Second, in metaphorical terminology, ground refers to a stabilizing, unchanging backdrop – the "bedrock" of humans' intentional action. From Descartes and Kant onward, the Earth has acted as the ground in both senses: the modern subject understands itself as a self-sovereign, reflective and agential being on the basis of, and in distinction from, the image of Earth as a stable, unchanging ground. Life distinguishes itself from non-life, and human life from non-human life, on the basis of intentional action: the inert Earth is the ground on which dynamic and vital life move. For Elizabeth Povinelli (2016), this division institutes an instrumental relation between life and non-life. It grounds an anthropocentric vision of non-life as given to human use, in which the inert becomes dynamic only as an external human agency molds it in accordance with its will.

The Anthropocene destabilizes this ground, for it gives the lie to any sense that Earth might be the stable and inert backdrop of human affairs. While it may appear to blur the division between humans and nature in a way that gives humans even *more* agency – now extended to the planetary scale – it *also* implies that human activity has forced the Earth system out of the Holocene's relatively stable climate patterns, and thus exposed human life to the very real possibility of its own extinction. The Anthropocene compels us to think in terms of Earthly forces that operate at timescales and magnitudes that defy human comprehension, and which have the capacity to create the ultimate catastrophe – not simply the end of life, but the end of the only life yet capable of self-conscious thought. It thus signals the return of the Kantian sublime. And just as Kant struggled with the arbitrariness of the Lisbon earthquake, so too does the Anthropocene confront us with the Earth's ultimate *indifference* to human life (Clark 2010). Rather than the central figure of planetary history, the Anthropocene positions humanity as a chance occurrence that can be wiped away as quickly as it appeared.

Clark's argument radically repositions both resilience and our sense of ethics and politics in the Anthropocene. To start, if we recognize that we owe a debt we can never repay to those humans and non-humans who came before us, then we can begin to recognize that our own actions will leave our descendants with certain kinds of worlds that are more or less habitable. Ethics and politics in the Anthropocene involve thinking through what kind of world we will bequeath to those who come after us. This parallels Simon Dalby's (2009, 2013) insistence that the world has become the object rather than the background of politics. In the era of the Anthropocene, the question becomes what kind of planet our politics will create.

The Anthropocene thus opens up a unique problem for thought and action: *even as we are threatened with extinction, we have the possibility of creating new life-worlds out of an increasingly ruined present*. As Box 7.2 discusses, the Anthropocene expresses a world-deforming potential that pushes the twentieth century's sense of indeterminacy beyond its limits. The Bauhaus, cyberneticians, neoliberal political economists, and early resilience thinkers all grappled with understanding and governing life in the absence of transcendental determinants of meaning and value, but they tended to approach these problems against the backdrop of a stable Earth where the possibility of thought itself was not called into question. They all acted within a world that was more or less inert, an unproblematic backdrop for dynamic human technology, culture, politics, and productive activity. But the Anthropocene calls this world into question: it makes the world itself the stakes of technology, culture, politics and production. The Anthropocene thus poses a new problem only tangentially related to sustainability. If sustainability raised the question of designing institutions that could ensure continued access to and use of environmental resources in both the present and future, then the Anthropocene calls into question the possibility of "using" resources in the first place. It thus radicalizes the question of indeterminacy that resilience scholars confront: in the Anthropocene, human activity creates the very world in which we live.

Geographers and scholars in cognate fields have been at the forefront of thinking through the implications this radical form of indeterminacy holds for environmental politics and ethics. Like resilience approaches, they mobilize forms of topological thought, but they tend to do so in ways that orient thought to the outside, to the possibilities for becoming-otherwise within and through the world. A few examples from geography and anthropology will draw out the difference.

To start, Sarah Whatmore's (2013) work on ontological politics draws on science and technology studies, political ecology and cultural geography to explore how the force of the non-human, manifest in disturbance events such as flooding, can destabilize seemingly transparent ontologies of a nature–society binary and the forms of scientific expertise that hinge on and perpetuate this ontology. While these events might unsettle established social, ecological and technical forms, the emergence of alternatives hinges on experimental engagements with knowledge production that can re-define and re-distribute expertise. This is a far more radical form of collaboration on complex problems than resilience offers, for it calls into question the status and expertise of scientific knowledge – including, we might add, knowledge on resilience. Importantly, Whatmore (2013) points to the importance of "slowing down" scientific research and dwelling on the objects that mediate different forms of knowledge on these events – such as photos of past floods that call into question the effectiveness of expert-driven flood mitigation strategies, or models with built-in errors that lead to well-intentioned but problematic flood mitigation initiatives. Slowness allows new forms of collaborative knowledge on the source of problems and potential solutions to emerge in a way that facilitates rather than forestalls new kinds of human–non-human entanglements. For Whatmore, new worlds – new affective arrangements between people and things, which generate new desires, capacities, meanings and values – can *potentially* emerge out of insecurities as disturbances provoke people to reflect, individually and collaboratively, on the quality of relations between each other and the non-human world, *and* the technical and material objects that mediate these relations. However, this worlding is a fraught process that hinges on the ability of experimental engagements to effectively de-center expertise and pave the way towards new ways of engaging with both the human and the non-human.

Jamie Lorimer's (2015) engagement with re-wilding experiments in Amsterdam's Oostvaardersplassen (OVP) likewise points to new practices of conservation science that knowingly experiment with uncertainty. Like ecological resilience approaches, managers in the OVP do not attempt to impose command-and-control-style practices. They essentially treat the OVP as a space where different species can interact, emerge, thrive, and/or suffer in more or less isolation from human interference. This is a form of conservation science that is open to uncertainty and surprise and does not attempt to impose order on the OVP. However, at the same time, this means it does not impose conventional conservationist meanings and values on the surprises that arise. This has led to controversies, such as conservationists publicizing covertly filmed videos of animals clearly ill-suited to the OVP's ecosystem suffering and dying (Lorimer and Driessen 2013). But it also illustrates forms of conservation science that do not fully interiorize potentiality. Rather than foreclosing the potentiality of an uncertain and indeterminate world, it allows surprises to unfold, follows them wherever they might lead, and thus produces new hybrid natures formed through multi-species agencies – even if these might not be the forms of nature conservationists themselves would value.

This geographic work on scientific knowledge controversies suggests that resilience is only *one* way of approaching and apprehending the Anthropocene's radical indeterminacy. Both Whatmore and Lorimer point to new forms of scientific practice that do not attempt to encode expressions of social and ecological difference in terms of adaptation-maladaptation and abstract-concrete. Instead, they describe practices of experimental learning that exceed the narrow functional concerns that dominate adaptive management and adaptive governance. And in both cases, the problem is not to enhance sustainability and secure meta-systemic stability, but rather to create new forms of social and ecological life in the face of uncertainty, indeterminacy and insecurity.

Research in cognate fields pushes these insights into world-making even further. Radical feminist and Indigenous scholars have considered the question of world-making on the threshold of catastrophe for some time, particularly in anthropology. Work on the anthropology of everyday violence has long explored the strategies and practices people use to live with the destabilizing effects of everyday terror (e.g., Taussig 1991), the inescapable potential for violence, death and loss (Scheper-Hughes 1992), and the unending trauma of past violence (Das 2007), to name but a few. This body of work forms the academic context for Mbembe and Roitman's (1995) passage that opens this chapter, and laid the conceptual foundations for more recent anthropological examinations of precarity and marginalization. For example, Anna Tsing's (2015) studies of matsutake mushroom picking campsites in Oregon raises the question of how life persists in eroded, destroyed, and otherwise ruined spaces. Tsing is sensitive to the pervasive insecurity of life on the margins of global liberal rule, and particularly to the way this life is untethered from linear, modernist temporal imaginaries of progress and development. Here, the durative and affective force of persisting drives creative engagements with ruined landscapes – but this creates possibilities for both new forms of sociability *and* capital accumulation. The matsutake trade that stretches from remote campsites in Oregon forests to exclusive Tokyo restaurants gradually abstracts the mushrooms from their context and folds them into global circuits of accumulation, even as these mushrooms sustain alternative forms of life within the campsites.

Elizabeth Povinelli's (2011a, 2016) research into what she calls the "economies of abandonment" that structure liberal life explicitly connects these concerns to political-philosophical debates over potentiality. Her work with indigenous Australian friends shows that Aboriginal struggles to persist within and outside of liberal rule are foreshortened by banal events that slowly erode potentiality: a washing machine breaks down and cannot rid clothes of debilitating parasites; neoliberal reforms undermine Aboriginal land claims, and so forth. In her reading, potentiality is always embedded in cruddy, enervating situations that sit at the interstice of different, often competing world-making projects. Importantly, she situates liberalism as one world-making project amongst others. Liberal worldings hinge on structuring relations between humans and the non-human world in a way that enframes the world as market values, and fashions subjects who can survive and thrive in this world – precisely the entrepreneurial, risk-taking subjects of neoliberal security rationalities. But this process of world-making makes the world increasingly *unlivable* for others. Liberalism creates what she calls a *demanding environment*: it makes a space for difference through cultural recognition, but demands difference make itself legible in some ways rather than others (Povinelli 2011a). We have seen this demanding environment in the demands resilience theory places on social and ecological difference to provide some kind of functional utility for multi-disciplinary efforts working to synthesize pragmatic sustainability solutions. In Povinelli's work, this demanding environment is particularly evident in Australian land rights cases following the Land Reform Act of 1976. The state made provisions to award legal title to peoples it determined were the traditional Indigenous owners of the land. But this determination hinged on certain assumptions about indigeneity itself: particularly, that Indigenous owners could demonstrate an unchanging totemic relation to the land. However, as Povinelli demonstrates, this demand effectively bracketed Indigenous strategies of survival during the (still ongoing) process of colonial invasions and forced relocations.

Povinelli's arguments demonstrate that the violence of liberal rule does not simply lie in the police state's exercise of sovereign power against those marginalized peoples whose being is marked as a threat to its order – although this is an important dimension to be sure. It also unfolds through banal practices of abandonment exercised on those whose suffering is deemed unworthy of collective, political response, or even perversely *valued* as a necessary sacrifice in the name of progress. This ethical orientation hinges on liberal practices of what she calls *geontopower* (Povinelli 2016). Geontopower describes the techniques and strategies through which liberal rule engages with difference *through* the division between Life and Non-Life. Liberal thought itself enframes the Non-Living as inert, and the Living as those entities, human and non-human, that possess vital and self-organizing metabolic capacities. This division grounds (in Clark's (2010) sense; see Box 7.2) liberal ethics: living existents are those who can potentially be the object of human responsibility and care; the non-living is given to instrumental use. It also becomes the matrix through which other non-liberal forms of life are judged: valued human life is that which is "properly" capable of distinguishing the living from the non-living; forms of life that cannot make this distinction are rendered as "living fossils," relics from the pre-historic past who have not yet become modern and civilized. Povinelli thus shows how world-making is a *perilous* process: the act of forming worlds suitable for some lives leaves others exposed to suffering, violence and death. And the demanding environment of recognition that attempts to account for those lives can often amplify rather than assuage these hardships.

These are essential arguments for both proponents and critics of resilience to think through. For the former, Povinelli's work cautions against the rush to synthesize expressions of social and ecological difference into pragmatic sustainability solutions, for the demanding environment of recognition can crush the force of becoming-otherwise. For the latter, her argument is a controversial caution against romanticizing potentiality. Against the tendency of some strands of critical theory to place unwavering faith in life's inventiveness, she shows that becoming-otherwise is dangerous: it places lives outside established forms of measure and value, and thus always at risk of violence or abandonment. Moreover, those who live as liberalism's other, such as her Indigenous friends, are often not the academics celebrating potentiality's life-affirming force. Povinelli's analysis thus opens the ethical problem of *whose* lives can and should bear this burden of suffering to make difference viable.

Taken together, research from geographers and anthropologists into world-making presents a problem that is qualitatively distinct from resilience thinking's concern with sustainability: it opens onto unanswerable ethical and political questions surrounding what forms of collective existence we might create out of present uncertainties, dislocations and insecurities, and how these forms of existence might (re)distribute the suffering, violence and hardships of becoming-otherwise. This research compels us to act, but also recognizes that action will always impact others, sometimes in negative ways. World-making is a call to heed human and non-human others, and to subject the categories, goals, techniques, rationalities and technologies that mediate these relations to the question of what kinds of voices and visibilities they facilitate – and what kinds of silences and invisibilities they create.

These are questions that belie tractable solutions. Instead, they are problems in a Deleuzian sense. For Deleuze, problems show how conventional framings of a given issue (such as sustainability) as having a determined series of possible solutions (true or false; more or less cybernetic adaptability) cannot capture the affective potentiality of bodies and discourses juxtaposed with one another. Problems thus direct thought to the outside: to the potential for becoming-otherwise. They provoke thought and spark conceptual innovations that potentially *transgress* the given order of things. As a Deleuzian problem, world-making in the Anthropocene cannot be reduced to a stepwise search for satisficing sustainable outcomes. Instead, it generates a demand for new ways of thinking and engaging indeterminacy. One possibility lies in cultural studies and international relations research into *hacking*.

HACKING THE ANTHROPOCENE

David Chandler's (2016b) recent work on big data and resilience shows how resilience-based governance strategies increasingly adopt the language and ethos of hacking. While this term is often associated with computer code, Chandler and cultural studies scholars such as McKenzie Wark (2004) suggest that hacking is a more general style of engaging with the world, which Wark argues can be traced back at least to the dawn of modern European capitalism. For Chandler (2016b: 4), a hack is "a form of intervention, which seeks to reveal new relations and interconnections." Hacking is a style of engagement that radicalizes the abstraction of systems thinking. As the Invisible Committee (2014: 43, emphasis in original) suggests in important language,

> whereas the engineer would capture everything that functions, in such a way that everything functions better in service to the system, the hacker asks himself "How does that work?" in order to find its flaws, but also to invent other uses, to experiment. Experimenting then means exploring what such and such a technique implies *ethically*. The hacker pulls techniques out of the technological system in order to free them.

Hacking is thus intimately concerned with systematicity and functionality. The precondition of hacking is understanding how something works as a system: what elements compose a system; and what relations between elements enable the system to function in some ways rather than others? But as the Invisible Committee emphasizes, it does not study a system in order to make the system function better. Rather, it studies the system in order to learn where exploits lie (Galloway 2007). These are points where the system might be subversively re-purposed and transformed, through its own systemic functioning, into something new. For example, in computer programming, exploits are points where code might be re-configured to make a program function in a different way. Hacking thus signals a creative activity of radical abstraction. It plays with the current arrangement of things, and the functional capacities that result from these arrangements, for no purpose other than freeing these capacities for use. As Wark (2004) argues, hacking generates the difference that makes difference.

As a form of ethical engagement with a complex world, hacking is always attuned to the immeasurable potential that saturates the present – and, importantly, the systemic structures that condition this potential and keep it in check. In this sense, we might think hacking as one expression of what Felix Guattari (2015a, 2015b) calls *transversalizing* practice. Transversality is a forerunner to Deleuze and Guattari's (1987) concept of *assemblage*, or *agencement*, which refers to heterogeneous arrangements of bodies, actions, desires, plans, statements, and laws that form "provisional contingent wholes," apparent unities always on the verge of dissolution (Pugh and Grove 2017). Assemblage and transversality both direct attention to the affective relations that saturate the present, and the immeasurable, immanent potential to become otherwise they express. But if the recent interest in assemblage theory can often be rightly criticized for romanticizing potentiality and privileging ontology over politics (Buchanan 2015), transversality directs attention to the techniques, technologies and practices that structure these affective relations *as* they bind us to the given order of things. It thus directs our attention to the inextricable affective entanglements of life and power. As Gary Genosko writes, transversality is "an element of militant practice that aims at a rupture with inherited modes of organization"; it gestures towards a micro-politics of "militant, social, responsive creativity" (Genosko 2002: 96, 104). This should not be confused with a crude, organicist vitalism that naively positions life before power. The de-structuring potentialities of transversality lie in those social, technical, and cultural machines that shape our subjectivity. Thus, our power and capacity to resist power, to generate new forms of life and new measures of value, comes from the way in which we are made subjects by the myriad assemblages that comprise capitalist order.

Transversality is a singular force, contextually specific and practical rather than an ontological category. Importantly, Guattari reads capitalism as a semiotic system that encodes, subjugates, and enslaves, human and non-human life. With the cybernetic term *enslavement*, Guattari emphasizes how capitalism operates through a variety of assemblages that machinically de- and re-code material flows as identities and values given meaning in relation to a transcendental master signifier of exchange value. The social and cultural machines through which economic production occurs thus *also* produce subjectivities more or less appropriate for insertion into these machines (Guattari 1992, 1996; Lazzarato 2014). While this may enable capitalist production to appropriate and exploit life on an increasingly intimate level of affective relations, it also enriches our collective potentialities in ways that threaten to exceed governmental regulation. The social, technical, and cybernetic machines that produce capitalist subjectivity also enable us to *potentially* engage with the world in new ways, and thus create new forms of life through and against mechanisms of subjection and control.

Guattari's arguments anticipated and, in many cases, influenced the analyses of world-making from Povinelli, Tsing, Whatmore, Lorimer and others. In general terms, we might read transversality as the world-(de) forming force of world-making. But his semiotically oriented sensibility explicitly positions world-making in relation to techniques of cybernetic regulation that shape contemporary capitalist subjectivities. In short, the concept of transversality directs our attention to interlocking matrices, the specific array of machinic assemblages that at once molds materialities and enunciations into extensive, delineated and individuated bodies and spaces – the demanding environments of specific life-worlds discussed above – and the intensive, affective potential for these individuated bodies and spaces to become other than they are.

Guattai's concern with transversalizing the machinic assemblages that concretize capitalist order helps us recognize how practices such as hacking, and the environmental techniques they mobilize to abstract and functionally re/combine systemic elements, can *both* be a vector for enhanced regulation *and* generate new capacities for resisting this control. Indeed, liberal rule targets hacking's capacities to create difference for appropriation, exploitation and, in some cases, extermination. Wark (2004) reads the history of capitalist development as a process of elite class appropriation of the hacks marginalized peoples create to survive and live dignified lives within and against capitalism's exploitative and dehumanizing practices of rule. More recently, Chandler (2016b) demonstrates how "digital hacktivism" in Jakarta transforms policymaking and security governance. As a governance strategy, hacking plays with big data and social media to transform threats and problems into systemic pressures that reveal latent interconnections and enable new forms of agency and community capacity.

At stake in hacking is the transversalizing capacity to generate difference through abstraction. As Wark (2004) describes it, to abstract is to "construct a plane upon which otherwise different and unrelated matters may be brought into many possible relations. To abstract is to express the virtuality of nature." This is a more radical, less stratified and less encoded form of abstraction than the abstraction that resilience initiatives attempt to produce. Hacking does not abstract in order to interiorize the outside. It does not attempt to abstract in ways that make concrete phenomena amenable to regulation through environmental techniques. Any interiorization is the result of capture, encoding and appropriation, whether this is through alienation and accumulation or the cybernetic will to design-in sustainability and adaptive capacity. Instead, hacking pushes capture beyond its limits: it abstracts in order to break free from the processes and relations that sustain the way things are. It is abstracting within and against context. As Alexander Galloway (2007) notes, hacking abstracts to remove impediments to development, change and becoming – to return being to its virtual, transversalizing structure, oriented towards the outside and the possibilities of becoming this opens onto.

Thus, even as it is vulnerable to capture, hacking works against the cybernetic abstraction that underpins resilience. If resilience thinking abstracts in order to systematize and disrupt management, hacking abstracts in order to de-systematize and render control ineffective. It is one way of ethically engaging with the abstract, virtual potentiality that resilience approaches designate as both threat and solution to sustainability – and it engages in a way that always leaves a remainder, a possibility for re-hacking and re-purposing that never exhausts the potential to become otherwise in a cybernetic will to design. Indeed, hacking the design process can transversalize this will to design itself, and re-purpose design as a method for ethical world-making. The next section explores how radical design scholars have begun to invent new forms of design practice.

TRANSITION DESIGN

Chapter 5 detailed how Herbert Simon's influence on ecologists installed a will to design at the heart of ecological resilience theory. As we saw above, a will to design animates a host of practices designed to make the self, communities, regions, nation-states, socio-technical infrastructure systems, and even the entire planet adaptive in the face of uncertainty and insecurity – even if these practices can work at contradictory cross-purposes to one another (see Chapter 2).

However, this Simonian spin is myopic and partial. Simon's particular slant on design emerged out of the specific way he visualized the decision-making process as a cybernetic interface between interior and exterior environments. Resilience scholars picked up and subtly tweaked the concepts Simon developed – especially hierarchy and bounded rationality – to create a science of ecological artifice that recalibrated truth and control around the pragmatic principles of adaptive management, adaptive governance, and institutional design. But this is only *one* way designerly sensibilities might inform the study of human–environment relations. Indeed, design studies scholars themselves are quick to emphasize that Simon's work is only one strand in a much broader and rapidly maturing academic field. Simon's cybernetic vision of design gained traction in the discipline – and continues to hold some sway today – because he offered a *scientistic* understanding of design at a time when the field was struggling to assert its relevance and independence. Even though Simon circulated within the same aesthetic regime of modernist design the Bauhaus consolidated, the latter's emphasis on the particularities of craftwork and learning through apprenticeship contrasted with Simon's own drive to develop a totalizing, all-encompassing theory of problem-solving in thinking beings, whether these were human, animal or machine.[1] His "science of design" thus provided some design scholars with a scientific definition of their field as the nomothetic science of decision-making processes in conditions of indeterminacy, rather than the idiographic practice of crafting artifice that the Bauhaus had practiced.

Thus, design studies scholars locate Simon as a key figure situated at a particular juncture in the discipline's history: as part of a wave of scientifically oriented researchers who sought to develop an account of design as a *rational process* of creating artifacts, even if rationality in a complex world means adaptively learning from a complex environment (Cross 2007; Bousbaci 2008; Johansson-Sköldberg *et al.* 2013). And while Simon's vision of design continues to resonate across the field today, other theoretical perspectives have emerged to offer competing visions of what design is and how it can be mobilized.

For example, Donald Schön defined his influential vision of a post-rational designer in explicit contrast to Simon's view of design as an adaptively rational problem-solving process. While he may overlook the more critical components of Simon's thought, and overstate Simon's adherence to optimization methods (see Meng 2009), Schön (1983: 14) also saw the designer as immersed in what he called "situations of practice" characterized by "complexity, uncertainty, instability, uniqueness, and value conflicts." These are not "problems to be solved," but are rather "problematic situations characterized by uncertainty, disorder and indeterminacy" that require "the active, synthetic skill of designing a desirable future and inventing ways of bringing it about" (ibid.: 15–16). This is a constructivist view of design as an "artistic, intuitive practice" (ibid.: 49) that both mobilizes and critiques tacit knowledge in order to redefine problematic situations. Importantly, Schön recognizes that reflection allows the designer to break free from established ways of knowing and addressing problems:

> through reflection, he [sic] can surface and criticize the tacit understandings that have grown up around the repetitive experiences of a specialized practice, and can make new sense of the situations of uncertainty or uniqueness which he may allow himself to experience. (ibid.: 61)

Schön thus shifts the analytical focus of design studies away from Simon's cybernetic *process* of problem-solving to the cognitive and contextualized *practices* designers deploy to make some kind of sense out of indeterminacy. Here, his arguments parallel Nigel Cross's (1982, 1999, 2007) work on design as a distinct style of thought. For both Cross and Schön, design names a certain attitude or comportment towards the world, an attunement to the limits of the present and the possibilities for creating new meanings and values out of the current arrangement of things.

Similarly, Richard Buchanan's (1992) influential theory of placements (discussed in Chapter 6) both draws on and departs from Simon's vision of design. Buchanan follows Schön, Cross and others who foreground the inventive and (designerly) critical nature of design practice. In this light, he argues that Simon's science of the artificial conflates a "science of existing human-made products whose nature Simon happens to believe is a manipulation of material and behavioral laws of nature" and "an inventive science of design thinking" (Buchanan 1992: 18–19). He is adamant on this point:

> [Simon] does not capture the radical sense in which designers explore the essence of what the artificial may be in human experience. This is a synthetic activity related to indeterminacy, not an activity of making what is undetermined in natural laws more determinate in artifice. (ibid.: 18)

Buchanan thus offers his theory of placements as a way to theoretically specify the tools and techniques designers use to bring their ethical comportment to bear on an indeterminate world. Placements enable designers to "intuitively or deliberately shap[e] a design situation, identifying the views of all participants, the issues of concern, and the intervention that becomes a working hypothesis" (Johansson-Sköldberg *et al.* 2013: 125). And in contrast to Simon's vision of problem-solving as an adaptive, stepwise process, this view of placements allows "the problem formulation and solution [to] go hand in hand rather than as sequential steps" (ibid.).

To be sure, this designerly sensibility has become an essential component of post-modern capitalism, which celebrates the near-infinite plasticity of being and the seemingly endless possibilities for self-design in a world unmoored from transcendental determinants of meaning and value, such as God or nature (Margolin 1995; Groys 2008). Jean Baudrillard (1981) highlighted precisely this complicity in his scathing critique of the Bauhaus, design and cybernetics. But in contrast to Baudrillard's scorched-earth tendencies, recognizing the situatedness and partiality of Simon's vision of design, and the alternative schools of designerly thought and practice that have emerged out of critical engagement with Simon, can help us recognize alternative methods for world-making and hacking the present.

Radical design scholars such as Tony Fry, Ezio Manzini and the Transitions Design scholars associated with the Carnegie Mellon School of Design (Irwin *et al.* 2015) have advanced pointed critiques of and, importantly, radical alternatives to conventional design theory and practice. For example, Fry (1999, 2017) frames the contemporary sense of unrelenting social, environmental and technological crises as a problem of *defuturing*. With this term, Fry signals how contemporary modes of living and interacting with both human and non-human others erode the constitutive relations on which living itself depends. In essence, defuturing is the paradoxical condition that results when a form of life premised on carboniferous capitalism as the engine of limitless growth and endless mass consumption actively negates the future – and the critical faculties for engaging with this negation (Fry 2010). He deploys this concept as a means of "confronting and removing the authority of the foundations of thought, upon which the narratives of the like of 'world,' 'future,' 'production' and 'progress' stand – this in order to make things otherwise" (Fry 1999: 2). And of course, design has been implicitly or explicitly complicit in defuturing since at least the Bauhaus, if not the colonial era (see Fry 2017; Tlostanova 2017). But Fry is interested in conceptual innovations that can re-purpose design and mobilize it to combat this defuturing interiorization. He argues for understanding design as an *ontological* activity that crafts the nature of beings that come into existence. We can get a sense of the ontological nature of design when we realize how thoroughly lives in the capitalist West have been designed: as Deleuze (1995b) emphasizes in his writings on control societies, the routines and rhythms that compose the ebb and flow of everyday life, our habits, intuitions and senses, our desires, aspirations and visions of what constitutes the "good" life are the effects of various environmental, biopolitical, security and sovereign techniques that strategically modulate the flow of bodies and the affective atmospheres through which they move. Here, design names the field of practice and mode of thought that has become formalized over the past century in response to the social, cultural, political and economic dislocations modernity has provoked.

Fry's work on defuturing and ontological design thus foregrounds the ethical and political dimensions of design practice. It repositions the problem of sustainability around design: if resilience scholars identify adaptability as both the source of and solution to problems of sustainability in a complex world, then Fry positions design as the source of and solution to the defuturing force of the Anthropocene. As he notes, ontological design is a recursive process: "all that is designed goes on designing" (2017: 26). This means that our artificial and designed modes of existence will continue to generate artificiality. While this might intensify the defuturing exploitation of human and non-human resources, this artificiality also gives a degree of indeterminacy – and thus plasticity – to foundational coordinates of our existence, such as our sense of the self, our understanding and practice of politics, and our modes of socialization. Taking seriously design's ethical and political dimensions thus enables us to recognize how it might contribute to radical projects for social and environmental justice and the crafting of alternative worlds.

Arturo Escobar (2018) stresses this point in his engagements with design studies. Escobar highlights points of critical juncture and disjuncture between indigenous social and environmental justice movements, particularly in South America, sustainability initiatives such as the Transition Towns Movement, and recent developments in radical design theory. He is careful to emphasize how all three engage with the problem of transition – although this is not simply the narrow vision of transition to sustainability that resilience proponents now emphasize. Instead, transition signals a broader activity of "reinventing the human" (Berry 1999, in

Escobar 2018) – creating new ways of living together with other humans and non-humans, a process akin to what we have been discussing as world-making. This is thus a far more radical sense of transition that opens onto alternative becomings not delimited by the cybernetic imperative to survive a catastrophic future by becoming resilient. Strategies such as re-localization, degrowth, or enshrining Rights of Nature into national constitutions all signal alternatives to the Western biopolitical project of growth, development and security through violently enforced property rights, instrumental use of non-human and human others, efficiency and profit maximization. Escobar focuses on the way two particular movements in design theory, Transitions Design and Design for Social Innovation, can inform these world-making efforts on both pragmatic and theoretical levels.

First, Transitions Design attempts to reconceptualize design practice and education in ways that facilitate the transition to alternative modes of living (Irwin *et al.* 2015). It departs from the recognition that "whilst our societies are in crisis, these crises are not being, and will never be, experienced in sufficiently motivating ways" (Tonkinwise 2015: 85–86). It thus mobilizes design techniques and strategies to plan for and manage social change in the absence of an eventful (in Povinelli's (2011a) sense) catastrophe that provokes a collective response. To be sure, Transitions Design draws heavily on the Transition Towns Movement, which aspires to create low-carbon economies by reorienting economic and cultural practices around localized and regionalized networks and alternative energy sources (Aitken 2012; Wilson 2012). However, it spins this through complexity science to develop an understanding of social and ecological change in terms of self-organizing, adaptive, decentralized and hierarchically organized networks. It thus risks amplifying the apolitical tendencies that characterize much Transition Town work (see Mason and Whitehead 2013) – and systems theory's de-historicized, apolitical and functionalist analysis of social dynamics, a point Escobar (2018) and Damian White (2015a) both stress. But at the same time, it also points beyond this resilience framing by explicitly expanding the possibilities of design practice. Specifically, it seeks to make designers (and design more broadly) accountable for the spatial and temporal implications of the artifacts they create. That is, it offers one way of concretizing Fry's (1999) call for an attunement to design's ontological implications: Transitions Design seeks to understand and consider how the practice of design draws together (or distances) human and non-human (organic and inorganic) elements across spatial distance and long-term time horizons in a way that makes certain kinds of worlds and forecloses others.

Ezio Manzini's (2015) recent work on *design for social innovation* similarly radicalizes design practice in a way that ethically engages the problem of world-making. Manzini sets his vision of design in opposition to the alienated detachment of modern design – both the detachment of the designer from the places or contexts in which artifacts are designed, produced and consumed, and the detachment implied in Simon's vision of a universally valid problem-solving process. Instead, Manzini emphasizes design as a place-maker. By this, he means that designers are increasingly taking "the local" as an object of design practice, as the Transition Design movement exemplifies. This "interest in places" (Manzini 2015: 44) immerses designers within the worlds they create, breaking down the cultural barriers that segmented designers, artifacts, and context of design. It also blurs the divide between expert designers and laypersons. Indeed, Manzini's sense of design as a place-making practice departs from the recognition that, in contemporary Western culture, *everybody designs*. This echoes Fry's sense of ontological design discussed above, which recognizes how subjects constituted through designerly techniques and strategies in turn design life-worlds that perpetuate this designerly mode of being. For Manzini, taking ontological design seriously leads to a new kind of designerly practice: not the top-down, expert-led participatory activities and charrettes that typify conventional design approaches, but rather "a *social conversation* in which different actors interact in different ways (from collaborating to conflicting) and at different times (in real time or off-line)" (ibid.: 49, emphasis in original). This is a collaborative co-designing process that works across difference in order to place the social itself under question: what kinds of forms of sociality might be invented as designers work with local peoples, experts, and other interested parties within particular environmental settings to invent new production systems, new ways of provisioning social services such as housing, new ways of producing and distributing food, and so forth?

In essence, Manzini describes the kinds of place-based collaborations Sarah Whatmore (2013) and colleagues developed to reconfigure flood management, as detailed above – but Manzini extends and intensifies the scope of these collaborations to encompass the activities and social and ecological systems that comprise everyday life itself. This is not a revolutionary, eventful change in everyday life, but rather a way of working through and across social and ecological difference to design for change in a way that preserves the potential for future change. In other words, it is a form of thought and practice that remains oriented towards the force of the outside. Indeed, Manzini's vision of design for social innovation takes seriously the ethical and political dimensions of world-making. He views collaborative place-making as a process that hinges on participants' active involvement, and thus emphasizes the importance of perpetuating the conditions for continued involvement. As Escobar (2018) emphasizes, his concern here parallels indigenous peoples' efforts to create a space for alterity in the face of modernity's debilitating homogenizing pressures. Design for social innovation thus offers one handle on the question of how world-making might proceed in a manner that takes account of the demanding environments and often unforeseen forms of suffering and violence world-making enacts on social and ecological difference: Manzini attempts nothing less than transforming design into a cross-boundary activity of creatively experimenting with sociality in ways that continually re-create the conditions for creative experimentation in the first place.

These brief discussions of radical design theorists do not do justice to the theoretical nuance, empirical details and practical experience that inform their work. Instead, it gestures towards work that critiques and transgresses design studies' cybernetic influence. Indeed, the wide breadth of this work, and its parallels with current research trends in geography, anthropology, and cognate fields, speaks to the *narrow* and *limiting* vision of design that underpins resilience approaches. With its conceptual roots in Simonian design theory, resilience is a socio-ecological expression of *one* form a will to design might take. But a will to design can inform other projects that do not deploy abstraction and functional synthesis in order to interiorize the force of the outside. Indeed, a designerly form of critique infuses the work of all these scholars, but they *also* recognize how design has been, and continues to be, a key mechanism in late modern capitalist subjection. The challenge, as they see it, is to work *through* design to fabricate new modes of living together. Even though the immensely resilient de-centralized capitalist order may appropriate these critical innovations (Hardt and Negri 2000), they nonetheless demonstrate that designerly sensibilities can inform radical alternative world-making projects that work at cross-purposes to resilience theory's interiorizing tendencies.

UN-WORLDING RESILIENCE

In this chapter, I have argued that resilience approaches create meaningful worlds as they draw together or hold apart social and ecological phenomena and give them meaning, value and identity through the series Sustainability–Institutional Design–Learning. Each point in this series is a zone of capture where topological thought interiorizes expressions of social and ecological difference, and makes becoming-otherwise become useful. Critical scholars provide a variety of different terms to help us conceptualize this process. For Brad Evans and Julian Reid (2014), this is where resilience *forgets how to die*; for Jeremy Walker and Melinda Cooper (2011), this is where resilience *metabolizes countervailing forces and inoculates itself against critique*; and for Tony Fry (1999), these are zones of *defuturing*. The series Anthropocene–Hacking–Transitions Design suggests ways that these zones of capture might be transformed into vectors that open outwards towards new becomings.

First, in contrast to resilience approaches, critical thought on the Anthropocene does not shoehorn the indeterminacy of world-making into the determined matrix of adaptation-maladaptation. Instead, it shows how world-making assembles embodied and embedded historical trajectories, aligning some with others in ways that realize some potentialities while foreclosing others. But it also reminds us that liberal rule often captures, suffocates, or exterminates this potentiality (Povinelli 2011a). It thus presents a unique ethical and political challenge: how to engage with the Anthropocene in ways that are attuned to the potential violence any world-making efforts might enact on those already striving to live beyond the measure of resilience (Grove 2017).

Second, hacking offers one strategy for engaging this challenge. As a technique for repurposing and transversalizing systematicity, hacking does not follow institutional design's cybernetic imperative to functionally synthesize diverse bounded rationalities around a pre-given problem, such as sustainability. Instead, hacking seeks to understand systematicity in order to play with, or transversalize the affective capacities environmental techniques generate. Hacking is thus abstraction within and against the world: it abstracts in order to inhabit and render monstrous the systemic arrangements that comprise contemporary forms of life. It constantly strives to generate difference out of the environmental techniques that modulate and interiorize difference.

Finally, radical design provides a suite of techniques and strategies for transversalizing those environmental techniques that interiorize the force of the outside. Radical design scholars do not shy away from design in light of the field's defuturing complicity within modern consumerist culture. Instead, much like Guattari encourages us to engage with and transversalize technologies of subjectivation, radical designers transversalize design itself. If we live in an increasingly artificial world in which everyone designs, what constitutive possibilities might this new reality hold? How might we learn to design the world differently, alongside human and non-human others, in ways that challenge rather than reinforce modernity's defuturing tendencies? And how might we design new forms of life that make a space that enhances rather than erodes the social, ecological, material, and technological conditions for continued experimentation with new ways of living? Transitions Design and designing for social innovation all suggest some ways designerly practices, techniques and strategies might contribute to this project of building worlds that open onto the possibilities for other worlds.

Thus, rather than capturing and ensnaring difference, the series Anthropocene–Hacking–Transitions Design deploys topological thinking in a way that turns thought and practice towards the outside. This is not a binary opposition – as scholars working on these themes emphasize, each category is a site where the political economic status quo can re-consolidate and reassert itself. But they are *also* points where difference might be engaged as difference and not subjected to the functionalist demand to become abstract and adaptive. There is thus no clear and straightforward solution to the possibilities for topological thought beyond resilience, for these possibilities are perilously perched on the threshold between interiorization and exteriorization. Moreover, many techniques and rationalities that might enable us to think the force of social and ecological difference in the Anthropocene are *also* those that interiorize this force in the name of building resilience and surviving the looming catastrophe. Thinking topologically beyond resilience thus requires a certain ethical comportment: an attunement to the demanding environment that research on human–environment relations creates, and the pressures this environment places on expressions of social and ecological difference. This barometric attentiveness can encourage scholars to question how they sense and engage with affective potentialities to become otherwise, and thus undo the defuturing worldings resilience initiatives enact.

FURTHER READING

Scholars in geography and allied fields are at the forefront of critical scholarship that thinks on and through the Anthropocene to re-envision alternative socio-ecological futures. Simon Dalby's *Security and Environmental Change* easily remains the most accessible introduction to the Anthropocene and its impacts on political thought. Similarly, Nigel Clark's *Inhuman Nature: Sociable Life on a Dynamic Planet* is a challenging but very rewarding exploration of how the Anthropocene destabilizes the grounds of Western subjectivity, philosophy and politics. And Elizabeth Povinelli's *Geontologies: A Requiem for Late Liberalism* uses the Anthropocene to tie together her wide-ranging body of work that stretches across anthropology, legal studies, continental philosophy, queer theory and critical indigenous studies – to name but a few – and examine the possibilities for ethics and politics in a groundless world. Alexander Galloway's work has long been at the forefront of critical cultural studies work on hacking; his texts *Protocol: How Control Exists after Decentralization* and *The Exploit: A Theory of Networks* (co-authored with Eugene Thacker) offer excellent introductions to debates in this discipline on cybernetic and networked forms of regulation and resistance. Finally, Ezio Manzini's *Design, When Everybody Designs: An Introduction to Design for Social Innovation* provides an accessible entry into state-of-the-art thinking in critical design studies.

NOTE

1 These divisions paralleled the debates between Richard Hartshorne and Fred Schaefer in geography during the 1950s. As is well documented within the discipline (Gregory 1978; Johnston 1979; Cox 2014), Schaefer contended that for geography to survive as a discipline it had to adopt the scientific method. Schaefer's call for a nomothetic, or generalizing, law-producing approach to geography as a spatial science contrasted with the idiographic, or particularizing approach to Hartshorne's regional geography, which dominated the field at the time.

8

CONCLUSIONS: RE-DESIGNING RESILIENCE?

INTRODUCTION

Given its ubiquity across academic and professional realms, there is seemingly no escaping resilience. It saturates thought and practice on living with uncertainty, and resilience proponents' aspirations to provide a universal account of social and ecological change leave precious little room for other ways of visualizing the forces that shape contemporary life. As the introductory chapter explained, this book has been concerned with the question of what resilience *does*. In general terms, I have attempted to show how resilience theory reconfigures the study of human–environment relations around a will to design that makes truth and control function in new ways: truth increasingly involves the production of rational abstractions amenable to functional synthesis; and control operates through systemic environmental engineering, or the cybernetic regulation of system performance through feedback-driven adaptive management strategies. But the genealogy I develop here also attempts to show that this is decidedly *not* a coherent or totalizing shift: recognizing the designerly roots of resilience situates resilience as *one way amongst others of understanding and responding to a generalized sense of crisis and insecurity* that has increasingly permeated Western societies since the end of the Cold War. As Chapter 2 indicated, this insecurity has been driven by unconventional threats such as environmental change, international terrorism, global economic turbulence, global narcotics and arms trafficking, and global pandemics give the lie to modernist security strategies premised on borders, separation and purification (Dalby 2002, 2009; Braun 2007; Dillon 2007). It has also been fueled by the failure of the past seventy years of modernist development to fulfill its promises of universal growth, prosperity and security. Seen in this light, resilience is one response to the insecurities that have been produced through, not in spite of, the historical trajectory of global capitalist development.

As this trajectory amplifies precariousness and insecurity in all facets of everyday life, the past decade has seen a proliferation of work on resilience. Across the academic disciplines and professional fields, individuals and communities in both the global North and South are enjoined to become resilient in the face of surprise and the loss of certainty. Critical scholars have been quick to draw attention to the depoliticizing effects of these interventions. But as the previous chapter detailed, the designerly roots of resilience exceed the categories and techniques of conventional critique. While much critical scholarship goes some way towards denaturalizing the truth-claims of resilience scholars, it often struggles to register resilience theory's designerly qualities that transform key categories of critical thought, such as conflict and local knowledge, into mechanisms of cybernetic control. Resilience gives politics and critique a certain kind of utility for researchers and managers, and it thus amplifies the need for both in order to drive forward the adaptive governance process. Conflict and critique becomes a mechanism for providing astute managers with feedback on system performance, which they can fold into revised governance initiatives. In political-philosophical terms, resilience thinking attempts to harness the dynamic and creative force of change, the force of becoming-otherwise, and *use* this force to create sustainable social and ecological systems in a complex and indeterminate world. As it recalibrates the study and management of human–environment relations around a will to design, resilience attempts to make becoming-otherwise become useful.

The genealogy of resilience traced in Chapters 2 through 5 thus opens onto a politics of potentiality. The designerly sensibility that implicitly motivates resilience scholarship compels scholars and practitioners alike to demand social and ecological difference present itself in ways that are useful for widespread efforts to synthesize sustainability solutions. This demand effectively interiorizes the force of the outside, or the potential to become otherwise. If we recognize that the topological interiorization of potentiality is ultimately what resilience does, then the challenge confronting geographers and scholars in allied fields is to return to the scene of the crime, so to speak – the terrain of social and ecological indeterminacy on and against which resilience operates – in order to invent new ways of engaging with the indeterminate and sublime potentiality that saturates our Anthropocene present.

This book has thus attempted to analyze the concept of resilience in a way that enables critical and applied scholars alike to engage with the concept in ways that move beyond its constituent limits. Specifically, the genealogy offered here has drawn attention to the designerly roots of resilience, the paradoxical practices of critique resilience theory mobilizes, and the politics of potentiality that follow on from this transformation. Each is reviewed in turn.

THE DESIGNERLY ROOTS OF RESILIENCE

Chapter 3 detailed how resilience scholars mobilize a topological sensibility attuned to systemic interconnection, non-linear emergence, and multiple-equilibrium systems to visualize resilience as a process of de- and re-formation, or changing form and function while maintaining identity. Topological thinking enabled resilience scholars to visualize systems as having the capacity to exist in multiple equilibrium states: because system performance is determined by qualitative relations between key elements, multiple configurations of these elements could provide the same level of output or service (Holling 1973). Rather than focusing analytical attention on individual component parts of a system, resilience scholars emphasized the importance of understanding how these components relate to one another. This "science of the integration of parts" (Holling 1995), which we have called here a "science of ecological artifice" to acknowledge Herbert Simon's influence, challenged many presuppositions of positivist science, and thus enabled resilience scholars to advance new techniques of adaptive management and adaptive governance for understanding and managing these systems.

Chapter 5 demonstrated how these topological insights drew heavily on Herbert Simon's cybernetic sciences of the artificial. Simon's specification of artifice as the topological interface between interior and exterior, and design as the practice of crafting artifice, infused ecological resilience with a will to design. Rather than reducing the world to objective truths that can be known and controlled with predictive certainty, resilience approaches attempt to reduce the world to rational abstractions that can be functionally synthesized with one another in order to develop pragmatic solutions to complex problems. Here, we can identify the designerly roots of resilience thinking, which can be traced back at least to the aesthetic transformations consolidated through (although by no means inaugurated by) the Bauhaus movement's engagements with early twentieth-century European crises. The Bauhaus promulgated a modernist design aesthetic, which apprehended the world as nothing more than abstract functions that could be endlessly re/combined with one another. As Jean Baudrillard (1981) details, this aesthetic underpinned post-World War II cyberneticians' grasp of life as information exchange, which in turn informed Simon's cybernetic spin on behavioral science.

Indeed, the diagrammatic qualities of Simon's thought emerge out of this modernist design aesthetic. He strove to apply an ontology of hierarchically organized complexity and boundedly rational individuals, an epistemology of simulation and an ethic of a will to design to *all* social and ecological phenomena – which the Bauhaus and cybernetics grasped in essence as signs denoting functionality. Simon thus deployed his diagrammatic understanding of complexity to create a certain kind of sense out of indeterminate and intractable problems in a number of disciplinary fields. Resilience theory extends this diagram to the study of human–environment relations. It advances a vision of social and ecological phenomena as complex, emergent and unpredictable, and provides ways of knowing and comporting oneself in relation to this complex world.

CRITIQUE FOR AND AGAINST THE PRESENT

Simon's influence on ecological resilience theory did not only reconfigure the study of human–environment relations around a will to design. A designerly sensibility also enabled resilience scholars to advance searing critiques of command-and-control management practices. Chapter 4 detailed how resilience proponents argued that managing for efficiency and optimization reduces diversity, inhibits flexibility, erodes adaptive capacity, and thus makes social and ecological systems susceptible to disturbance and collapse (Holling and Meffe 1996). In contrast, they advocated for new styles of environmental management based on the principles of adaptive management and, later, adaptive governance. Adaptive management attempts to build feedback-driven learning into the management process, while adaptive governance attempts to transform on the broader cultural and institutional context in which management occurs. In both cases, the point is to generate new institutions that mediate between complex systems and their environment – whether this narrowly targets science–management relations, or wider cultural, political and economic practices, respectively.

Critical scholars often latch onto these programs of cultural engineering and institutional change as examples of the ideological affinities between resilience initiatives and neoliberal governance reform. And there is much to these critiques, to be sure. As Chapter 2 detailed, resilience programming often focuses on developing individual and systemic capacities to survive and thrive in the face of vulnerability and insecurity. Rather than avoid risk, resilience proponents assert that risk and disturbance can be a source of change, innovation, growth and adaptation. Much like neoliberal governance, resilience initiatives thus have the biopolitical effect of fashioning entrepreneurial subjects that embrace risk. Here, risk management becomes individualized and depoliticized: rather than a result of historically specific unequal and unjust political economic relations, vulnerability and insecurity are an inescapable part of the world that individuals are more or less capable of coping with. Those who cannot cope with risk become marked for biopolitical programs of cultural engineering and institutional change designed to build their adaptive capacities and increase resilience.

But as Chapter 6 suggested, there is a more complex relation between resilience and critique. The designerly roots of resilience inscribe a particular *designerly* critical ethos at the heart of resilience theory. Resilience initiatives do not necessarily seek to demonstrate the partiality of truth claims, or reveal how power relations shape what we commonly take for granted. Rather than attempting to destabilize the present, resilience attempts to *transform* the present through experimentally (re)combining elements with one another, playing with elements' functional attributes in ways that generate new identities, meanings and functions out of the existing arrangement of things. This means that there is no optimal solution to building resilience. Instead, it involves an indeterminate process of experimental interventions that "tinker" with (Ostrom 1999), "play" with (Wylant 2010; Buchanan 1992), or "recombine" (Folke *et al.* 2007) elements that comprise a system.

This designerly style of critique allows resilience to both work with and, potentially, against the ongoing neoliberalization of political ecological relations. The influence of both pragmatist political philosophy and neo-liberal public choice theory infuse resilience-based adaptive governance initiatives with a designerly critical ethos, which asserts that experimenting with different institutional arrangements can empower marginalized peoples and give voice to alternative ways of knowing and interacting with the environment, that conventional environmental governance arrangements often sideline. However, this is also a cybernetic form of empowerment, which frames empowerment as a techno-managerial process of removing institutional, political and cultural barriers to communication, deliberation, and information flow more broadly. Similarly, the adaptive management initiatives early resilience scholars championed sought to transform the environmental management process into a stepwise, adaptive search for effective policy interventions. In contrast to top-down, optimized, one-size-fits-all solutions, adaptive management involves a process of tinkering, or experimenting with different policy interventions in a way that allows scientists and managers to visualize how systems responded to disturbance and adjust their intervention on the basis of this new knowledge gained. Of course, this model is rarely realized in practice, but it demonstrates the importance resilience scholars place on institutional change – in this case, the need to recalibrate the formal and informal rules, norms and procedures that determine how information flows between scientists, managers, and complex social and ecological systems.

This means that resilience is at once *more* neo-liberal than many critiques acknowledge *and* potentially more politically pliable than many critics allow. In terms of the former, resilience theory gained traction during the mid-1990s as they took up and modified neo-liberal public choice theorists' work on institutional design and constitutional choice in relation to problems of sustainability. Public choice theory enabled resilience scholars to transform their critique of command and control into a tractable problem of institutional fit, and transform conflict into a mechanism of cybernetic control (see Chapter 4). However, these two strands of work converge around a shared modernist design aesthetic that grasps human and non-human existents as nothing more than rational abstractions amenable to functional synthesis. While this aesthetic gave rise to cybernetics and novel techniques of control and biopolitical regulation (Baudrillard 1981; Deleuze 1995b; Hardt and Negri 2000), it has also informed more radical political projects that strive to create a space for social and ecological difference.

The alignment between neoliberalism and resilience is thus far more contingent than many critics allow (see also Chandler 2014a; Schmidt 2015). Resilience radically transforms environmental management to the extent that it foregrounds relationality and the affective potential that inheres the human and non-human alike. In many ways, it signals the triumph of the post-1968 left to advance a vision of governance not premised on centralized regulation (Chandler 2014a). And while its underlying cybernetic sensibilities often mobilize environmental techniques in ways that capture difference and make the force of becoming-otherwise become useful, it also gestures towards a suite of critical practices and techniques that might inform other critical projects striving to create more just socio-ecological futures. At issue is how we might engage this embodied, affective potentiality in ways that destabilize and subvert, rather than cybernetically reinforce, prevailing socio-ecological orders.

THE POLITICS OF POTENTIALITY

Recognizing the designerly roots of resilience and its paradoxical relation to critique suggests that we cannot simply dismiss resilience as nothing more than the latest fold in neoliberal environmental governance. Instead, resilience opens onto an indeterminate politics of potentiality. While Kai Lee (1999: unpag.) might argue that adaptive management "recapitulates the promise that Francis Bacon articulated four centuries ago," we might rather suggest that resilience theory recapitulates the problem that confronted early twentieth-century philosophers, managers, planners, and artists: how to orient themselves in relation to a radically indeterminate future that cannot be tamed through modernist security techniques such as prediction or territorializing boundaries. Resilience is one way of getting a handle on indeterminacy – and its convergence with neo-liberal public choice theory has made it the state-sanctioned response to indeterminacy (Duffield 2011) – but others are surely possible. At stake is how we might relate – individually and collectively – to the potential to become otherwise that saturates an indeterminate present. For Brad Evans and Julian Reid (2014), resilience thinking compels us to turn from this potential – what they call in philosophical terms "forgetting how to die" – and cling to the promise of surviving the catastrophe a bit longer by becoming resilient. Transforming the present here often amounts to little more than changing in ways that ensure the present arrangement of things may persist. And indeed, resilience scholars recognize as much when they define resilience as the topological capacity of people, societies and ecosystems to "be subjected to disturbance and cope, without changing their 'identity' – without becoming something else" (Walker and Salt

2012: 3). Coping here is a form of topological security (Grove 2017) that transforms the experience of indeterminacy – the wavering of the present that signals how things might become other than they are – into a stabilizing force.

But research in geography and allied disciplines shows that there are other ways we might relate to the experience of difference and indeterminacy. Chapter 7 described how work in geography and anthropology on world-making in the Anthropocene introduces new ethico-political concerns that de-center the privilege resilience thinking gives to future systemic (in)security and sustainability. World-making in the absence of transcendental determinants of meaning and value is an ethico-aesthetic practice that forces us to engage with the actually existing, embodied socio-ecological difference that surrounds us all, even if liberal governance techniques often blind us to their striving. Research in cultural studies and international relations on hacking indicates a variety of techniques and attitudes for subversively re-purposing cybernetic governance techniques. Hacking resilience holds out the promise of engaging in forms of affective engineering that do not interiorize affective potential. Similarly, radical design studies points to a number of initiatives that already take seriously the need to work through and against resilience techniques in order to design more equitable and just socio-ecological futures.

Importantly, these different research threads also mobilize a variety of theoretical concepts that allow us to re-envision the challenge of sustainability in ways that sidestep a resilience framing and return to the founding problem of how to engage indeterminacy. Tony Fry's sense of *defuturing* helps us recognize how modern governance techniques – including, we should note, resilience – draw our attention, resources and capacities towards ensuring *future* systemic meta-stability while eroding our capacity for apprehending our current situation and addressing the pervasive insecurities and vulnerabilities that saturate our *present*. Felix Guattari's concept of *transversality* helps us recognize that the environmental techniques that instantiate cybernetic forms of socio-ecological governance might also endow us with capacities to engage in programs of affective engineering that combat and re-purpose rather than solidify modern capitalist political ecological order. Nigel Clark's concept of *un-grounding* heightens our sense of responsibility in the Anthropocene to produce a world that is habitable for both the humans and non-humans who follow after us. And Elizabeth Povinelli's concept of *geontologies* attunes us to the ethical problem world-making opens: any radical project that strives to re-purpose socio-ecological relations – no matter how well-intentioned – will necessarily impinge on other world-making projects, and potentially increase hardship and suffering for those already struggling to persist as difference.

Thus, resilience lies at the heart of a politics of potentiality that revolves around the question of what kind of world we might design in the Anthropocene – and how we might go about designing this world, and who should bear the burden of transformation. Even though it is embroiled within this politics, resilience also has much to offer: it already mobilizes a variety of critical sensibilities and environmental techniques calibrated to engage with socio-ecological indeterminacy. The challenge for critical scholarship moving forward is devising ways to hack resilience – to abstract the concept and its techniques from their cybernetic origins and thus enable novel forms of affective engineering that combat rather than enhance the interiorization of potentiality. This is a key point we might take from the genealogy offered here: as a product of (one strand of) designerly thinking, resilience theory is nothing more than an artifact itself. And as an artifact, resilience theory can thus potentially be re-designed around a different set of ethical and political commitments. Can geographers and allied scholars across the sciences, humanities, and design fields re-design resilience into a concept that addresses already-existing insecurities rather than future systemic insecurities?

DESIGNING RESILIENCE IN THE ANTHROPOCENE

This book has thus unpacked how the concept of resilience has influenced thought and practice in geography and cognate disciplines as it moved out of isolated corners of ecology (and, to a lesser extent for our purposes here, psychology and engineering) and became a key organizing principle in fields such as environmental management, disaster management and development, homeland security, urban planning, climate change adaptation and critical infrastructure protection, to name but a few. Resilience theory recalibrates the study of human–environment relations around a will to design, which offers a number of threats and opportunities. On one hand, a will to design animates environmental techniques that enable resilience scholars and practitioners to visualize, abstract and act on the affective relations that make up everyday life. Resilience approaches effectively delimit how researchers can know and engage with social and ecological change: they fold expressions of social and ecological difference into a cybernetic understanding of sustainability, and attempt to build individual and systemic capacities to live with and adapt to change in order to survive and thrive within an insecure and catastrophic future. Resilience approaches thus often further entrench neoliberal governance strategies that seek to naturalize insecurity and develop individual risk-taking capacities. But on the other hand, they also introduce critical sensibilities into social and environmental governance. They depart from the recognition that centralized solutions often create unforeseen consequences, and provide a powerful call for more inclusive governance arrangements. While this often results in post-political, technocratic approaches to vulnerability and insecurity, it also has the potential to make politics and critique function in new ways. Resilience thus provides opportunities – however limited in practice – to transform (rather than revolutionize) state–society relations and make the institutions of modern life more responsive to the needs of vulnerable and marginalized peoples.

This bifurcation means that any effort to ascribe a normative value to resilience will inevitably silence one of these tendencies. Resilience cannot be uncritically adopted as the solution to challenges of social and ecological complexity, for this ignores how resilience has become a vital cog in the ongoing neoliberalization of social and ecological relations. But at the same time, resilience cannot simply be dismissed as a post-political concept, for this ignores how resilience approaches have transformed governance in critical ways that radical scholars are still developing theoretical tools to fully appreciate.

What to do with resilience then? In many ways, resilience has become a problem in a Deleuzian sense: a provocation to thought that demands sustained and intensive *engagement*. What might engagement with resilience entail? First, it requires more sustained and detailed explorations of how resilience approaches operate in practice. This is one of the lessons we can draw from radical scholarship on hacking: opening systems, and forms of systematicity, to the potential to become otherwise first requires a nuanced understanding of how the system operates itself. In this light, critical scholars' knee-jerk dismissals of resilience as nothing more than a cover for ongoing neoliberalization actively *deter* the kinds of knowledge needed to understand how resilience approaches operate in particular contexts and in response to particular forms of problems (Anderson 2015; Simon and Randalls 2016). Instead, engaging with resilience might involve more detailed ethnographic and comparative studies of resilience programs in operation: how do scholars and practitioners actually go about building resilience? How do they conceptualize and operationalize resilience in specific projects, and how do resilience initiatives complement or juxtapose with other initiatives? How do resilience initiatives take up, modify and re-purpose existing governance techniques and strategies, and how do the existing arrangements of people, things, technologies, and strategies confound and transform resilience efforts? It also might involve more detailed archival study into the roots of resilience thinking that builds on genealogies of resilience in disciplines such as computer science (McQuillan 2016; Cavanagh 2016), ecology and economics (Zebrowski 2016; Chandler 2014a; Evans and Reid 2014; Walker and Cooper 2011), and design (as this book offers). Importantly, this work could also inform ethnographic research on actually existing resilience in order to trace how these divergent roots continue to shape theory and practice in a variety of fields today. Taking the example of design from this book, the mode of engagement I am suggesting here might build on previous genealogical work to explicitly explore how resilience approaches redefine the process of identifying problems for academic study and policy solutions, the formulation of different strategies and methods for addressing these problems, and the practices of theoretical, epistemological and practical synthesis mobilized to develop solutions. Taken together, these engagements with resilience will go some way towards understanding how actually existing resilience functions in everyday practice – and this understanding is the basis for identifying points of transgression where critical interventions might repurpose resilience and direct it towards the cultivation rather than negation of radical alterity.

Recognizing resilience's designerly roots also suggests that engaging with resilience might entail more sustained engagements with design – both design studies as a distinct and potentially complementary field of integrative research and design as a distinct, synthetic mode of apprehending and engaging with indeterminacy. The links between critical social theory and design have yet to be fully elaborated, although there have been initial efforts in anthropology (Escobar 2018), international relations (Weber 2012; Weber and Lacy 2011) and environmental sociology (White 2015a, 2015b). But there is a glaring need for geographers in particular to engage with design. Many of the most difficult questions confronting geographers today, such as the status of the political in an interconnected and indeterminate world or the process of world-making in the Anthropocene, are questions designers have explored with rigorous theoretical and practical considerations. And designers have also developed nuanced critiques of Herbert Simon's cybernetic vision of design that could offer much to critical analyses of resilience theory. In many ways, design is ahead of geography and cognate fields when it comes to thinking about what to *do* in order to collaboratively build more socially and environmentally just futures. But at the same time, geographers have much to offer design as well. To start, geographers have advanced critiques of resilience that caution against some radical designers' more uncritical adoptions of elements of ecological resilience theory. And as an integrative discipline that often emphasizes the difference place makes (Massey 2005), geographers are well positioned to contribute their knowledge and skills to design's growing interest in holistic place-making. Indeed, geographers anticipated practices of re-localization that interest many radical designers today, such as J.K. Gibson-Graham's (2006) work on alternative economies or Sarah Whatmore and colleagues' place-based approach to flood management. In many ways, geography is ahead of design when it comes to understanding how place and space matter in contested processes of social and ecological change.

The concept of resilience thus presents geographers and allied scholars with new possibilities for broadening the scope of research into both human–environment relations and spatiality in an indeterminate world. While it extends and intensifies the possibilities for control, regulation and capture, it also opens onto designerly styles of thought and practice that could offer significant contributions to interdisciplinary and trans-boundary efforts to address the most challenging ethical and political problems that confront scholars today. Rather than turning from resilience, the challenge is to explore the different ways geographers and other scholars might creatively and collaboratively re-design resilience in the Anthropocene.

References

Adams, W.M. 2001. *Green Development: Environment and Sustainability in the Third World* (2nd ed.). London: Routledge.
Adey, P. 2008. Airports, mobility, and the calculative architecture of affective control. *Geoforum* 39: 438–451.
Adger, W.N. 2000. Social and ecological resilience: are they related? *Progress in Human Geography* 24: 347–364.
Agamben, G. 1978. *Infancy and History: Essays on the Destruction of Experience*. New York: Verso.
Agamben, G. 1990. *The Coming Community*. Minneapolis: University of Minnesota Press.
Agamben, G. 2005. *State of Exception*. Chicago: University of Chicago Press.
Agamben, G. 2007. *The Kingdom and the Glory: For a Theological Genealogy of Economy and Government*. Stanford, CA: Stanford University Press.
Agamben, G. 2009. *What is an Apparatus? And Other Essays*. Stanford, CA: Stanford University Press.
Agamben, G. 2014. What is a destituent power? *Environment and Planning D: Society and Space* 32(1): 65–74.
Agrawal, A. 2005. *Environmentality: Technologies of Government and the Making of Subjects*. Durham, NC: Duke University Press.
Aitken, G. 2012. Community transitions to low carbon futures in the transition towns network (TTN). *Geography Compass* 6(2): 89–99.
Allen, J. 2003. *Lost Geographies of Power*. Oxford: Blackwell.
Allen, J. 2011. Topological twists: power's shifting geographies. *Dialogues in Human Geography* 1(3): 283–298.
Allen, T. and Starr, T. 1982. *Hierarchy: Perspectives for Ecological Complexity*. Chicago: University of Chicago Press.
Amin, A. 2002. Spatialities of globalization. *Environment and Planning A* 34: 385–399.
Amin, A. and Thrift, N. 2002. *Cities: Reimagining the Urban*. London: Polity Press.
Anderies, J., Janssen, M., and Ostrom, E. 2004. A framework to analyze the robustness of social-ecological systems from an institutional perspective. *Ecology and Society* 9(1): 18.
Anderson, B. 2009. Affective atmospheres. *Emotion, Space and Society* 2: 77–81.

Anderson, B. 2010a. Preemption, precaution, preparedness: anticipatory action and future geographies. *Progress in Human Geography* 34(6): 777–798.
Anderson, B. 2010b. Modulating the excess of affect: morale in a state of 'total war.' In M. Gregg and G. Seigworth (eds), *The Affect Theory Reader*. Durham, NC: Duke University Press, pp. 161–185.
Anderson, B. 2012. Affect and biopower: towards a politics of life. *Transactions of the Institute of British Geographers* NS 37: 28–43.
Anderson, B. 2015. What kind of thing is resilience? *Politics* 35(1): 60–66.
Anderson, B. and Adey, P. 2011. Affect and security: exercising emergency in 'UK civil contingencies.' *Environment and Planning D: Society and Space* 29(6): 1092–1109.
Anderson, B., Kearnes, M., McFarlane, C., and Swanton, D. 2012. On assemblages and geography. *Dialogues in Human Geography* 2(2): 171–189.
Aradau, C. 2014. The promise of security: resilience, surprise and epistemic politics. *Resilience: International Policies, Practices, and Discourses* 2(2): 73–87.
Aradau, C. and van Munster, R. 2011. *Politics of Catastrophe: Genealogies of the Unknown*. London: Routledge.
Arendt, H. 1968. Introduction: Walter Benjamin, 1892–1940. In W. Benjamin, *Illuminations*. New York: Schocken.
Arrow, K., Bolin, B., Costanza, R., Dasgupta, P., Folke, C., Holling, C.S., Jansson, B.O., Levin, S., Maler, K., Perrings, C., and Pimental, D. 1995. Economic growth, carrying capacity and the environment. *Science* 268: 520–521.
Ball, H. 1974. Dada manifesto. In H. Ball, *Flight Out of Time: A Dada Diary*. Berkeley: University of California Press.
Barnett, C. 2005. The consolations of 'neoliberalism.' *Geoforum* 36(1): 7–12.
Barr, S. 1964. *Experiments in Topology*. Mineola, NY: Dover Publications.
Baudrillard, J. 1981. *For a Critique of the Political Economy of the Sign*. New York: Telos Press.
Beckford, G. 1972. *Persistent Poverty: Underdevelopment and Plantation Economies of the Third World*. Kingston: University of the West Indies Press.
Beilin, R. 2015. Introduction: governing for urban resilience. *Urban Studies* 52(7): 1205–1217.
Benjamin, W. 1968a. Theses on the philosophy of history. In W. Benjamin, *Illuminations*. New York: Schocken.
Benjamin, W. 1968b. The work of art in the age of mechanical reproduction. In W. Benjamin, *Illuminations*. New York: Schocken.
Berkes, F. 1999. *Sacred Ecology*. New York: Routledge.
Berkes, F. and Folke, C. 1998. Linking social and ecological systems for resilience and sustainability. In F. Berkes and C. Folke (eds), *Linking Social and Ecological Systems: Management Practices and Social Mechanisms for Building Resilience*. Cambridge: Cambridge University Press, pp. 1–25.
Berkes, F., Colding, J., and Folke, C. 2003. Introduction. In F. Berkes, J. Colding and C. Folke (eds), *Navagating Social-Ecological Systems: Building Resilience for Complexity and Change*. Cambridge: Cambridge University Press, pp. 1–29.
Berlant, L. 2011. *Cruel Optimism*. Durham, NC: Duke University Press.
Blaikie, P. and Brookfield, H. 1987. *Land Degradation and Society*. London: Methuen.
Bogues, A. 2002. Politics, nation, and postcolony: Caribbean inflections. *Small Axe* 11: 1–30.
Boltanski, L. 2011. *On Critique: A Sociology of Emancipation*. London: Polity.
Botkin, D. 1990. *Discordant Harmonies: A New Ecology for the Twenty-First Century*. Oxford: Oxford University Press.

Bousbaci, R. 2008. 'Models of Man' in design thinking: The 'bounded rationality' episode. *Design Issues*, 24: 38–52.
Bowker, G. 1993. How to be universal: some cybernetic strategies, 1943–70. *Social Studies of Science* 23: 107–127.
Boyd, E., and Juhola, S. 2009. Stepping up to the climate change: opportunities in re-conceptualizing development futures. *Journal of International Development* 21: 792–804.
Brand, F. and Jax, K. 2007. Focusing on the meaning(s) of resilience: resilience as a descriptive concept and a boundary object. *Ecology and Society* 12(1): 23.
Braudel, F. 1992. *The Structures of Everyday Life: Civilization and Capitalism, 15th-18th Century, Vol. 1*. Berkeley: University of California Press.
Braun, B. 2002. *The Intemperate Rainforest: Nature, Culture and Power on Canada's West Coast*. Minneapolis: University of Minnesota Press.
Braun, B. 2007. Biopolitics and the molecularization of life. *Cultural Geographies* 14(1): 6–28.
Brown, K. 2003. Integrating conservation and development: a case of institutional misfit. *Frontiers in Ecology and the Environment* 1(9): 479–487.
Brown, K. 2013. Global environmental change I: a social turn for resilience? *Progress in Human Geography* 38(1): 107–117.
Brown, K. 2016. *Resilience, Development and Global Change*. London: Routledge.
Brown, K. and Westaway, E. 2011. Agency, capacity, and resilience to environmental change: lessons from human development, well-being, and disasters. *Annual Review of Environment and Resources* 36: 321–342.
Brown, T. 2009. *Change by Design: How Design Thinking Transforms Organizations and Inspires Innovation*. New York: HarperCollins.
Brown, W. 1998. Genealogical politics. In J. Moss (ed.), *The Later Foucault: Politics and Philosophy*. London: Sage.
Bryant, R. and Bailey, S. 1997. *Third World Political Ecology*. London: Routledge.
Buchanan, I. 2015. Assemblage Theory and its Discontents. *Deleuze Studies* 9(3): 382–392.
Buchanan, J. 1959. Positive economics, welfare economics, and political economy. *The Journal of Law and Economics* 2: 124–138.
Buchanan, J. 1999. Natural and Artifactual Man. In *The Collected Works of James M. Buchanan, Volume 1: The Logical Foundations of Constitutional Liberty*. Indianapolis: Liberty Fund.
Buchanan, J. and Tullock, G. 1962. *The Calculus of Consent: The Logical Foundations of Democracy*. Indianapolis: Liberty Fund.
Buchanan, R. 1992. Wicked problems in design thinking. *Design Issues* 8(2): 5–21.
Bulkeley, H. and Castán Broto, V. 2013. Government by experiment? Global cities and the governing of climate change. *Transactions of the Institute of British Geographers* NS 38: 361–375.
Burton, I., Kates, R., and White, G. 1978. *The Environment as Hazard*. New York: Guilford.
Campbell, D. 1992. *Writing Security: United States Foreign Policy and the Politics of Identity*. Minneapolis: University of Minnesota Press.
Canguilhem, G. 1994. The Concept of Life. In F. Delaporte (ed.), *A Vital Rationalist: Selected Writings from Georges Canguilhem*. New York: Zone Books, pp. 303–320.
Cannon, T., and Mueller-Mahn, D. 2010. Vulnerability, resilience and development discourses in context of climate change. *Natural Hazards* 55: 621–635.

Castree, N. 2004. *Nature*. London: Routledge.
Cavanagh, C. 2016. Resilience, class and the antifragility of capital. *Resilience: International Policies, Practices, Discourses*. Online pre-print available at http://dx.doi.org/10.1080/21693293.2016.1241474, accessed 3 May 2017.
Chandler, D. 2013. Resilience and the autotelic subject: toward a critique of the societalization of security. *International Political Sociology* 7: 210–226.
Chandler, D. 2014a. *Resilience: The Art of Governing Complexity*. London: Routledge.
Chandler, D. 2014b. Beyond neoliberalism: resilience, the new art of governing complexity. *Resilience: International Policies, Practices, Discourses* 2(1): 47–63.
Chandler, D. 2016a. Resilience: The societalization of security. In D. Chandler and Julian Reid (eds), *The Neoliberal Subject: Resilience, Adaptation and Vulnerability*. London: Rowman and Littlefield, pp. 27–50.
Chandler, D. 2016b. Securing the Anthropocene? International policy experiments in digital hacktivism: A case study of Jakarta. *Security Dialogue* 48(2): 113–130.
Chester, D. 2001. The 1755 Lisbon earthquake. *Progress in Physical Geography* 25(3): 363–383.
Cinner, J. 2011. Socio-ecological traps in reef fisheries. *Global Environmental Change* 21: 835–839.
City of New York. 2013. *A Stronger, More Resilient New York*. New York: Office of the Mayor.
Clark, N. 2010. *Inhuman Nature: Sociable Life on a Dynamic Planet*. London: Sage.
Clark, N. 2014. Geo-politics and the disaster of the Anthropocene. *The Sociological Review* 62(S1): 19–37.
Coaffee, J. 2013. From securitisation to integrated place making: towards next generation urban resilience in planning practice. *Planning Practice and Research* 28(3): 323–339.
Coaffee, J. and Rogers, P. 2008. Rebordering the city for new security challenges: from counter-terrorism to community resilience. *Space and Polity* 12(1): 101–118.
Coaffee, J., Murakami Wood, D., and Rogers, P. 2009. *The Everyday Resilience of the City: How Cities Respond to Terrorism and Disaster*. New York: Palgrave Macmillan.
Cohen, D. 2016. Interviews with Rebuild by Design's Group of Experts: Henk Ovink. *Public Culture* 28(2): 317–327.
Collier, S. and Lakoff, A. 2008. The vulnerability of vital systems: How 'critical infrastructure' became a security problem. In M. Dunn, and K. Kristensen. (eds), *The Politics of Securing the Homeland: Critical Infrastructure, Risk and Securitization*. London: Routledge.
Collier, S. 2009. Topologies of power: Foucault's analysis of political government beyond 'governmentality.' *Theory, Culture & Society* 26(6): 78–108.
Collier, S. 2011. *Post-Soviet Social: Neoliberalism, Social Modernity, Biopolitics*. Princeton, NJ: Princeton University Press.
Collier, S. 2017. Neoliberalism and rule by experts. In V. Higgins and W. Larner (eds), *Assembling Neoliberalism: Expertise, Practices, Subjects*. New York: Palgrave.
Colwell, C. 1997. Deleuze and Foucault: series, event, genealogy. *Theory and Event* 1(2).
Connolly, W. 1993. *The Terms of Political Discourse* (3rd ed.). Princeton, NJ: Princeton University Press.
Cooper, M. 2010. Turbulent worlds: financial markets and environmental crisis. *Theory, Culture & Society* 27(2–3): 167–190.

Cote, M. and Nightingale, A. 2012. Resilience thinking meets social theory: Situating social change in socio-ecological systems (SES) research. *Progress in Human Geography* 36(4): 475–489.
Cox, K. 2014. *Making Human Geography*. New York: Guilford.
Cox, S. and Cox, P. 2016. *How the World Breaks: Life in Catastrophe's Path, from the Caribbean to Siberia*. New York: The New Press.
Coyne, R. 2005. Wicked problems revisited. *Design Studies* 26: 5–17.
Cretney, R. 2014. Resilience for whom? Emerging critical geographies of resilience. *Geography Compass* 8/9: 627–640.
Cross, N. 1982. Designerly ways of knowing. *Design Studies* 3(4): 221–227.
Cross, N. 1999. Design research: a disciplined conversation. *Design Issues* 15(2): 5–10.
Cross, N. 2007. Forty years of design research. *Design Studies* 28(1): 1–4.
Crowther-Heyck, H. 2005. *Herbert A. Simon: The Bounds of Reason in Modern America*. Baltimore, MD: The Johns Hopkins University Press.
Crutzen, P. 2002. Geology of mankind. *Nature*. 415: 23.
Dalby, S. 1991. Critical Geopolitics – Discourse, Difference, and Dissent. *Environment and Planning D–Society & Space* 9(3): 261–283.
Dalby, S. 2002. *Environmental Security*. Minneapolis: University of Minnesota Press.
Dalby, S. 2009. *Security and Environmental Change*. London: Polity.
Dalby, S. 2013. Biopolitics and climate security in the Anthropocene. *Geoforum* 49: 184–192.
Das, V. 2007. *Life and Words: Violence and the Descent into the Ordinary*. Berkeley: University of California Press.
Davidson, D. 2010. The applicability of the concept of resilience to social systems: some sources of optimism and nagging doubts. *Society and Natural Resources* 23: 1135–1149.
Davoudi, S. 2012. Resilience: a bridging concept or a dead end? *Planning Theory & Practice* 13(2): 299–333.
Dawley, S., Pike, A., and Tomaney, J. 2010. Towards the resilient region? *Local Economy* 25(8): 650–667.
Dean, M. 2004. *Governmentality: Power and Rule in Modern Society*. London: Sage.
deGoede, M. 2012. *Speculative Security: The Politics of Pursuing Terrorist Monies*. Minneapolis: University of Minnesota Press.
DeLanda, M. 1991. *War in the Age of Intelligent Machines*. New York: Zone Books.
DeLanda,M. 2002. *Intensive Science and Virtual Philosophy*. New York: Continuum.
DeLanda, M. 2006. *A New Philosophy of Society: Assemblage Theory and Social Complexity*. New York: Continuum.
Deleuze, G. 1988. *Foucault*. Minneapolis: University of Minnesota Press.
Deleuze, G. 1990. *Logic of Sense*. New York: Continuum.
Deleuze, G. 1995a. The exhausted. *SubStance* 24(3): 3–28.
Deleuze, G. 1995b. Control and becoming/Postscript on control societies. In G. Deleuze, *Negotiations*. New York: Columbia University Press.
Deleuze, G. 2004. How do we recognize structuralism? In G. Deleuze, *Desert Islands and Other Texts: 1953–1974*. Los Angeles: Semiotext(e), 2004, pp. 170–192.
Deleuze, G. and Guattari, F. 1987. *A Thousand Plateaus: Capitalism and Schizophrenia*. Minneapolis: University of Minnesota Press.
Demeritt, D. 1994. The nature of metaphors in cultural geography and environmental history. *Progress in Human Geography* 18(2): 163–185.

Derrida, J. 2009. *The Beast & the Sovereign, Volume 1*. Chicago: University of Chicago Press.
Dewey, J. 1927. *The Public and its Problems*. Athens, OH: Swallow Press.
Dewey, J. 1938. *Experience and Education*. New York: Touchstone.
DFID. 2011. *Defining Disaster Resilience: A DFID Approach Paper*. London: UK Department for International Development.
DHS. 2013. *NIPP 2013: Partnering for Critical Infrastructure Security and Resilience*. Washington, D.C.: US Department for Homeland Security.
Dietz, T., Ostrom, E., and Stern, P. 2003. The struggle to govern the commons. *Science* 302: 1907–1912.
Dillon, M. 1996. *Politics of Security: Towards a Political Philosophy of Continental Thought*. New York: Routledge.
Dillon, M. 2007. Governing through contingency: the security of biopolitical governance. *Political Geography* 26(1): 41–47.
Dillon, M. 2008. Underwriting security. *Security Dialogue* 39(2–3): 309–332.
Dillon, M. and Lobo-Guerrero, L. 2008. Biopolitics of security in the 21st century: an introduction. *Review of International Studies* 34: 265–292.
Dillon, M. and Reid, J. 2009. *The Liberal Way of War: Killing to Make Life Live*. London: Routledge.
Dittmer, J. 2010. Textual and discourse analysis. In D. DeLyser, S. Herbert, S. Aitken, M. Crang and L. McDowell, *The Sage Handbook of Qualitative Geography*. London: Sage.
Dreyfus, H. 1965. Alchemy and artificial intelligence. RAND Corporation, P-3244. December 1965.
Duffield, M. 2007. *Development, Security, and Unending War: Governing the World of Peoples*. Cambridge: Polity.
Duffield, M. 2008. Global civil war: the non-insured, international containment and post-interventionary society. *Journal of Refugee Studies* 21(2): 145–165.
Duffield, M. 2010. The liberal way of development and the development-security impasse: exploring the global life-chance divide. *Security Dialogue* 41(1): 53–76.
Duffield, M. 2011. Total war as environmental terror: linking liberalism, resilience, and the bunker. *The South Atlantic Quarterly* 110(3): 757–769.
Duffield, M. 2012. Challenging environments: danger, resilience, and the aid industry. *Security Dialogue* 43(5): 475–492.
Dunn-Cavelty, M., Kaufmann, M., and Kristensen, K. 2015. Resilience and (in)security: practices, subjects, temporalities. *Security Dialogue* 46(1): 3–14.
Eagleton, T. 1990. *The Ideology of the Aesthetic*. Oxford: Blackwell.
Eakin, H. and Leurs, A. 2006. Assessing the vulnerability of social-environmental systems. *Annual Review of Environment and Resources* 31: 365–394.
Elden, S. 2001. *Mapping the Present: Heidegger, Foucault, and the Project of a Spatial History*. New York: Continuum.
Elden, S. 2006. *Speaking Against Number: Heidegger, Language, and the Politics of Calculation*. Edinburgh: Edinburgh University Press.
Ernston, H., van der Leeuw, S., Redman, C., Meffert, D., Davis, G., Alfsen, C., and Elmqvist, T. 2010. Urban transitions: on urban resilience and human-dominated ecosystems. *Ambio* 39: 531–545.
Escobar, A. 2012. *Notes on the Ontology of Design*. Paper presented at the Sawyer Seminar, Indigenous Cosmopolitics: Dialogues about the Reconstitution of Worlds, organized by Marisol de La Cadena and Mario Blaser, October 30. University of California, Davis.

Escobar, A. 2018. *Designs for the Pluriverse: Radical Interdependence, Autonomy and the Making of Worlds*. Durham, NC: Duke University Press.
Esposito, R. 2011. *Immunitas: The Protection and Negation of Life*. Cambridge: Polity Press.
Evans, B. and Reid, J. 2014. *Resilient Life: The Art of Living Dangerously*. London: Polity.
Evans, J. 2011. Resilience, ecology and adaptation in the experimental city. *Transactions of the Institute of British Geographers* 36(2): 223–237.
Evans, J., Karvonen, A., and Raven, R. (eds) 2016. *The Experimental City*. London: Routledge.
Ewald, F. 1991. Insurance and risk. In G. Burchell, C. Gordon, and P. Miller (eds), *The Foucault Effect: Studies in Governmentality*. Chicago, IL: University of Chicago Press.
FEMA 2011. *A Whole Community Approach to Emergency Management: Principles, Themes, and Pathways for Action*. FDOC 104–008–1. Washington, D.C.: Federal Emergency Management Agency.
Folke, C. 2006. Resilience: the emergence of a perspective for social–ecological systems analyses. *Global Environmental Change* 16: 253–267.
Folke, C., Carpenter, S., Elmqvist, T., Gunderson, L., Holling, C.S., and Walker, B. 2002. Resilience and Sustainable Development: Building Adaptive Capacity in a World of Transformations. *Ambio* 31(5): 437–440.
Folke, C., Colding, J., and Berkes, F. 2003. Synthesis: building resilience and adaptive capacity in social-ecological systems. In F. Berkes, J. Colding, and C. Folke (eds), *Navigating Social-Ecological Systems: Building Resilience for Complexity and Change*. Cambridge: Cambridge University Press, pp. 352–387.
Folke, C., Hahn, T., Olsson, P., and Norberg, J. 2005. Adaptive governance of social-ecological systems. *Annual Review of Environment and Resources* 30: 441–473.
Folke, C., Pritchard, Jr., L., Berkes, F., Colding, J., and Svedin, U. 2007. The problem of fit between ecosystems and institutions: ten years later. *Ecology and Society* 12(1): 30.
Folke, C., Carpenter, S., Walker, B., Scheffer, M., Chapin, T., and Rockström, J. 2010. Resilience thinking: integrating resilience, adaptability and transformability. *Ecology and Society* 15(4): 20.
Foster, H. 2002. *Design and Crime (And Other Diatribes)*. London: Verso.
Foster, H. 2015. *Bad New Days: Art, Criticism, Emergency*. London: Verso.
Foucault, M. 1972. *The Archaeology of Knowledge*. New York: Pantheon.
Foucault, M. 1977a. Nietzsche, genealogy, history. In *Language, Counter-Memory, Practice*. Ithaca, NY: Cornell University Press.
Foucault, M. 1977b. Theatrum philisophicum. In *Language, Counter-Memory, Practice*. Ithaca, NY: Cornell University Press.
Foucault, M. 1989. *The Order of Things: An Archaeology of the Human Sciences*. New York: Vintage.
Foucault, M. 1990. *The History of Sexuality, Volume 1*. New York: Vintage.
Foucault, M. 1994. *The Birth of the Clinic: An Archaeology of Medical Perception*. New York: Vintage.
Foucault, M. 1995. *Discipline and Punish: The Birth of the Prison*. New York: Vintage.
Foucault, M. 1998a. A preface to transgression. In J. Faubion. (ed.), *Michel Foucault: Aesthetics, Method, and Epistemology*. New York: The New Press.

Foucault, M. 1998b. The thought of the outside. In J. Faubion (ed.), *Michel Foucault: Aesthetics, Method, and Epistemology*. New York: The New Press.
Foucault, M. 1998c. Return to history. In J. Faubion (ed.), *Michel Foucault: Aesthetics, Method, and Epistemology*. New York: The New Press.
Foucault, M. 2003. *'Society Must be Defended': Lectures at the Collège de France, 1975–1976*. New York: Picador.
Foucault, M. 2004. *Death and the Labyrinth*. New York: Continuum.
Foucault, M. 2007a. *Security, Territory, Population: Lectures at the Collège de France, 1977–1978*. New York: Picador.
Foucault, M. 2007b. What is Critique? In M. Foucault, *The Politics of Truth*. Los Angeles: Semiotext(e).
Foucault, M. 2008. *Birth of Biopolitics: Lectures at the Collège de France, 1978–1979*. New York: Picador.
Fry, T. 1999. *A New Design Philosophy: An Introduction to Defuturing*. Sydney: University of New South Wales Press.
Fry, T. 2010. *Design as Politics*. Oxford: Berg.
Fry, T. 2017. Design for/by 'The Global South.' *Design Philosophy Papers* 15(1): 3–37.
Gaillard, J.C. 2010. Vulnerability, capacity and resilience: perspectives for climate and development policy. *Journal of International Development* 22: 218–232.
Galison, P. 1994. The ontology of the enemy: Norbert Wiener and the cybernetic vision. *Critical Inquiry* 21(1): 228–266.
Gallie, W.B. 1956. Essentially contested concepts. *Proceedings of the Aristotelian Society* 56: 167–198.
Galloway, A. 2007. *Protocol: How Control Exists after Decentralization*. Cambridge, MA: MIT Press.
Garmestani, A. and Benson, M. 2013. A framework for resilience-based governance of social-ecological systems. *Ecology & Society* 18(1): 9.
Genosko, G. 2002. *Felix Guattari: An Aberrant Introduction*. New York: Continuum.
Gerlach, J. and Jellis, T. 2015. Guattari: impractical philosophy. *Dialogues in Human Geography* 5(2): 131–148.
Gibson-Graham, J.K. 2006. *A Post-Capitalist Politics*. Minneapolis: University of Minnesota Press.
Gordon, C. 1986. Question, ethos, event: Foucault on Kant and Enlightenment. *Economy and Society* 15(1): 71–87.
Gregory, D. 1978. *Ideology, Science and Human Geography*. London: Hutchinson.
Gregory, D., Johnston, R., Pratt, G., Watts, M., and Whatmore, S. 2009. *The Dictionary of Human Geography, 5th Edition*. Oxford: Wiley-Blackwell.
Grimm, N., Grove J., Pickett, S., and Redman, C. 2000. Integrated approaches to long-term studies of urban ecological systems. *Biosciences* 50: 571–584.
Grimm, N., Faeth, S., Golubiewski, N., Redman, C., Wu, J., Bai, X., and Briggs, J. 2008. Global change and the ecology of cities. *Science* 319: 756–760.
Gropius, W. 1919. *Bauhaus Manifesto and Program*. Weimar: The administration of the Staatliche Bauhaus.
Grosz, E. 2001. Architecture from the outside. In E. Grosz, *Architecture from the Outside: Essays on Virtual and Real Space*. Cambridge, MA: MIT Press.
Grosz, E. 2009. *Chaos, Territory, Art: Deleuze and the Framing of the Earth*. New York: Columbia University Press.
Grove, K. 2010. Insuring "our common future?" Dangerous climate change and the biopolitics of environmental security. *Geopolitics* 15(3): 536–563.

Grove, K. 2012. Preempting the next disaster: catastrophe insurance and the financialization of disaster management. *Security Dialogue* 43(2): 139–155.
Grove, K. 2013a. From emergency management to managing emergence: a genealogy of disaster management in Jamaica. *Annals of the Association of American Geographers* 103(3): 570–588.
Grove, K. 2013b. Hidden transcripts of resilience: power and politics in Jamaican disaster management. *Resilience* 1(3): 193–209.
Grove, K. 2013c. Biopolitics. In C. Death (ed.), *Critical Environmental Politics*. London: Routledge, pp. 22–30.
Grove, K. 2014a. Agency, affect, and the immunological politics of disaster resilience. *Environment and Planning D: Society and Space* 32: 240–256.
Grove, K. 2014b. Adaptation machines and the parasitic politics of life in Jamaican disaster resilience. *Antipode* 46(3): 611–628.
Grove, K. 2017. Security beyond resilience. *Environment and Planning D: Society and Space* 35(1): 184–194.
Grove, K. and Adey, P. 2015. Security and the politics of resilience: an aesthetic response. *Politics* 35(1): 78–84.
Grove, K. and Chandler, D. 2017. Introduction: resilience and the Anthropocene: the stakes of 'renaturalising' politics. *Resilience: International Policies, Practices and Discourses* 5(2): 79–91.
Grove, K. and Pugh, J. 2015. Assemblage thinking and participatory development: diagram, ethics, biopolitics. *Geography Compass* 9(1): 1–13.
Groys, B. 2008. The Obligation to Self-Design. *e-flux* #0: 1–7.
Groys, B. 2016. *In the Flow*. London: Verso.
Guattari, F. 1992. *Chaosmosis: An Ethico-Aesthetic Paradigm*. Sydney: Power Publications.
Guattari, F. 1996. Semiological subjection, semiotic enslavement. In G. Genosko (ed.), *The Guattari Reader*. Oxford: Blackwell, pp. 141–147.
Guattari, F. 2015a. Transversality. In F. Guattari, *Psychoanalysis and Transversality: Texts and Interviews 1955–1971*. Los Angeles: Semiotext(e).
Guattari, F. 2015b. *Lines of Flight: For Another World of Possibilities*. New York: Bloomsbury.
Gunderson, L. 2000. Ecological resilience – in theory and application. *Annual Review of Ecological Systems* 31: 425–439.
Gunderson, L. and Holling, C. (eds) 2002. *Panarchy: Understanding Transformations in Human and Natural Systems*. Washington, DC: Island.
Gunderson, L., Holling, C.S., and Light, S. (eds.) 1995a. *Barriers and Bridges to the Renewal of Ecosystems and Institutions*. New York: Columbia University Press.
Gunderson, L., Holling, C.S., and Light, S. 1995b. Barriers broken and bridges built: a synthesis. In L. Gunderson, C.S. Holling and S. Light (eds), *Barriers and Bridges to the Renewal of Ecosystems and Institutions*. New York: Columbia University Press, pp. 489–532.
Gunderson, L., Garmestani, A., Rizzardi, K., Ruhl, J., and Light, A. 2014. Escaping a rigidity trap: governance and adaptive capacity to climate change in the Everglades social ecological system. *Idaho Law Review* 51: 127–156.
Hacking, I. 1990. *The Taming of Chance*. Cambridge: Cambridge University Press.
Hagan, J. 1992. *An Entangled Bank: The Origins of Ecosystem Ecology*. New Brunswick, NJ: Rutgers University Press.
Hanna, S., Folke, C., and Maler, K-G. (eds) 1996. *Rights to Nature: Ecological, Economic, Cultural, and Political Principles of Institutions for the Environment*. Washington, DC : Island Press.

Haraway, D. 1988. Situated knowledges: the science question in feminism and the privilege of partial perspective. *Feminist Studies* 14(3): 575–599.
Haraway, D. 1997. *Modest_Witness@Second_Millennium FemaleMan_Meets_OncoMouse: Feminism and Technoscience*. London: Routledge.
Hardin, G. 1968. The tragedy of the commons. *Science* 162(3859): 1243–1248.
Hardt, M. and Negri, A. 2000. *Empire*. Cambridge, MA: Harvard University Press.
Harvey, D. 1973. *Social Justice and the City*. Baltimore, MD: Johns Hopkins University Press.
Harvey, D. 1985. The geopolitics of capitalism. In D. Gregory and J. Urry (eds), *Social Relations and Spatial Structures*. New York: Springer.
Harvey, D. 2005. *A Brief History of Neoliberalism*. New York: Oxford University Press.
Harvey, D. 2007. The Kantian roots of Foucault's dilemmas. In S. Elden and J. Crampton (eds), *Space, Knowledge, Power: Foucault and Geography*. London: Routledge.
Hayek, F. 1945. The use of knowledge in society. *The American Economic Review* 35(4): 519–530.
Heidegger, M. 1996. *Being and Time*. Albany: SUNY Press.
Hewitt, K. 1983. The idea of calamity in a technocratic age. In K. Hewitt (ed.), *Interpretations of Calamity from the Viewpoint of Human Ecology*. Boston: Allen & Unwin.
Heyck, H. 2008a. Defining the computer: Herbert Simon and the bureaucratic mind – part 1. *IEEE Annals of the History of Computing*. April–June 2008: 42–51.
Heyck, H. 2008b. Defining the computer: Herbert Simon and the bureaucratic mind – part 2. *IEEE Annals of the History of Computing*. April–June 2008: 42–63.
Hodson, M. and Marvin, S. 2009. Urban ecological security: a new urban paradigm? *International Journal of Urban and Regional Research* 33: 193–215.
Hollander, G. 2012. *Raising Cane in the 'Glades: The Global Sugar Trade and the Transformation of Florida*. Chicago: University of Chicago Press.
Holling, C.S. 1973. Resilience and stability of ecological systems. *Annual Review of Ecological Systems* 1973 4: 1–23.
Holling, C.S. (ed.) 1978. *Adaptive Environmental Assessment and Management*. New York: Wiley.
Holling, C.S. 1986. The resilience of terrestrial ecosystems: local surprise and global change. In W. Clark and R. Munn (eds), *Sustainable Development of the Biosphere*. Cambridge: Cambridge University Press.
Holling, C.S. 1992. Cross-scale morphology, geometry, and dynamics of ecosystems. *Ecological Monographs* 62: 447–502.
Holling, C.S. 1995. What barriers? What bridges? In L. Gunderson, C.S. Holling, and S. Light (eds), *Barriers and Bridges to the Renewal of Ecosystems and Institutions*. New York: Columbia University Press, pp. 3–34.
Holling, C.S. 1996. Engineering resilience versus ecological resilience. In P. Schulze (ed.), *Engineering within Ecological Constraints*. Washington, DC: National Academy Press.
Holling, C.S. 2001. Understanding the complexity of economic, ecological, and social systems. *Ecosystems* 4: 390–405.
Holling, C.S. 2004. From complex regions to complex worlds. *Ecology and Society* 9(1): 11.
Holling, C.S. and Dantzig, G. 1976. Determining Optimal Policies for Ecosystems. University of British Columbia, Institute of Resource Ecology Paper R-7-B.

Holling, C.S. and Goldberg, M. 1971. Ecology and planning. *Journal of the American Institute of Planners* 37(4): 221–230.
Holling, C.S. and Meffe, G. 1996. Command and control and the pathology of natural resource management. *Conservation Biology* 10(2): 328–337.
Holling, C.S. and Sanderson, G. 1996. Dynamics of (dis)harmony in ecological and social systems. In S. Hanna, C. Folke and K-G. Maler (eds), *Rights to Nature: Ecological, Economic, Cultural, and Political Principles of Institutions for the Environment*. Washington, DC: Island Press, pp. 57–86.
Holling, C.S., Gunderson, L., and Peterson, G. 2002. Sustainability and panarchies. In L. Gunderson and C.S. Holling (eds), *Panarchy: Understanding Transformations in Human and Natural Systems*. Washington DC: Island Press, pp. 63–102.
Huppatz, D.J. 2015. Revisiting Herbert Simon's "Science of Design." *Design Issues* 31(2): 29–40.
Hussain, N. 2003. *The Jurisprudence of Emergency: Colonialism and the Rule of Law*. Ann Arbor: University of Michigan Press.
Hutchings, J. and Myers, R. 1994. What can be learned from the collapse of a renewable resource? Atlantic cod, *gadus morhua*, of Newfoundland and Labrador. *Canadian Journal of Fisheries and Aquatic Sciences* 51: 2126–2146.
Hyndman, J. 2001. Towards a feminist geopolitics. *The Canadian Geographer* 45(July): 210–222.
The Invisible Committee. 2014. *To Our Friends*. Cambridge, MA: MIT Press.
Irwin, T., Kossoff, G., and Tonkinwise, C. 2015. Transition Design provocation. *Design Philosophy Papers* 13(1): 3–11.
Johansson-Sköldberg, U., Woodilla, J., and Çetinkaya, M. 2013. Design thinking: past, present and possible futures. *Creativity and Innovation Management* 22(2): 121–146.
Johnson, P. 2008. Foucault's spatial combat. *Environment and Planning D: Society and Space* 26: 611–626.
Johnston, R. 1979. *Geography and Geographers: Anglo-American Human Geography Since 1945*. London: Edward Arnold.
Jones, M. 2009. Phase space, relational thinking, and beyond. *Progress in Human Geography* 33(4): 487–506.
Jones, R., Pykett, J., and Whitehead, M. 2013. *Changing Behaviours: On the Rise of the Psychological State*. Cheltenham: Edward Elgar.
Kanngieser, A. 2013 *Experimental Politics and the Making of Worlds*. Aldershot: Ashgate.
Karvonen, A. and van Heur, B. 2014. Urban laboratories: experiments in reworking cities. *International Journal of Urban and Regional Research* 38(2): 379–392.
Kates, R. 1962. *Hazard and Choice Perception in Flood Plain Management*. Department of Geography Research Paper No. 78, Department of Geography, The University of Chicago.
Kaufmann, S. 1995. *At Home in the Universe: The Search for the Laws of Self-Organization and Complexity*. Oxford: Oxford University Press.
Kaufmann, S. 2000. *Investigations*. Oxford: Oxford University Press.
Kay, L. 2000. *Who Wrote the Book of Life? A History of the Genetic Code*. Stanford, CA: Stanford University Press.
Kemp, R. and Rotmans, J. 2005. The management of the co-evolution of technical, environmental and social systems. In M. Weber and J. Hemmelskamp (eds), *Towards Environmental Innovation Systems*. New York: Springer, pp. 33–55.

Kiser, L. and Ostrom, E. 1982. The three worlds of action: a metatheoretical synthesis of institutional approaches. In E. Ostrom (ed.), *Strategies of Political Inquiry*. London: Sage, pp. 179–222.

Klein, N. 2007. *The Shock Doctrine: The Rise of Disaster Capitalism*. New York: Picador.

Koehler, K. 2009. The Bauhaus Manifesto postwar to postwar: from the street to the wall to the radio to the memoir. In J. Saletnik and R. Schuldenfrei (eds), *Bauhaus Construct: Fashioning Identity, Discourse and Modernism*. New York: Routledge.

Koselleck, R. 1988. *Critique and Crisis: Enlightenment and the Pathogenesis of Modern Society*. Cambridge, MA: MIT Press.

Kuhn, T. 1962. *The Structure of Scientific Revolutions*. Chicago: University of Chicago Press.

Lakoff, A. 2007. Preparing for the next emergency. *Public Culture* 19(2): 247–271.

Lash, S. 2012. Deforming the figure: topology and the social imaginary. *Theory, Culture & Society* 29(4/5): 261–287.

Latour, B. 1993. *We Have Never Been Modern*. Cambridge, MA: Harvard University Press.

Latour, B. 2004. Why has critique run out of steam? From matters of fact to matters of concern. *Critical Inquiry* 30: 225–248.

Latour, B. 2008. A cautious Prometheus? A few steps toward a philosophy of design (with special attention to Peter Sloterdijk). Keynote lecture for the *Networks of Design* meeting of the Design History Society, Falmouth, Cornwall, 3 September 2008.

Lazzarato, M. 2006. From biopower to biopolitics. *Tailoring Biotechnologies* 2(2): 11–20.

Lazzarato, M. 2014. *Signs and Machines: Capitalism and the Production of Subjectivity*. Los Angeles: Semiotext(e).

Leach, M. and Mearns, R. (eds) 1996. *The Lie of the Land: Challenging Received Wisdom on the African Environment*. Portsmouth: Heinemann.

Lebel, L., Anderies, J., Campbell, B., Folke, C., Hatfield-Dodds, S., Hughes, T., and Wilson, J. 2006. Governance and the capacity to manage resilience in regional social-ecological systems. *Ecology and Society* 11(1): 19.

Lee, K. 1993. *Compass and Gyroscope: Integrating Science and Politics for the Environment*. Washington, DC: Island Press.

Lee, K. 1999. Appraising adaptive management. *Ecology and Society* 3(2): 3.

Lemert, C. and Gillan, G. 1982. *Michel Foucault: Social Theory and Transgression*. New York: Columbia University Press.

Lentzos, F. and Rose, N. 2009. Governing insecurity: contingency planning, protection, resilience. *Economy and Society* 38(2): 230–254.

Levin, S. 1992. The problem of pattern and scale in ecology. *Ecology* 73(6): 1943–1967.

Levin, S. 1998. Ecosystems and the biosphere as complex adaptive systems. *Ecosystems* 1: 431–436.

Light, S., Gunderson, L., and Holling, C.S. 1995. The Everglades: evolution of management in a turbulent environment. In L. Gunderson, C.S. Holling, and S. Light (eds), *Barriers and Bridges to the Renewal of Ecosystems and Institutions*. New York: Columbia University Press.

Loos, A. 1997. Ornament and crime. In A. Loos, *Ornament and Crime: Selected Essays*. Riverside, CA: Aradine Press.

Lorimer, J. 2015. *Wildlife in the Anthropocene: Conservation after Nature*. Minneapolis: University of Minnesota Press.
Lorimer, J. and Driessen, C. 2013. Wild experiments at the Oostvaardersplassen: rethinking environmentalism in the Anthropocene. *Transactions of the Institute of British Geographers* 39: 169–181.
MacKinnon, D. and Derickson, K. 2012. From resilience to resourcefulness: a critique of resilience policy and activism. *Progress in Human Geography* 37(2): 253–270.
MacLeod, G. 2002. From urban entrepreneurialism to a "revanchist city"? On the spatial injustices of Glasgow's renaissance. *Antipode* 34(3): 602–624.
Malabou, C. 2012. *Ontology of the Accident: an Essay on Destructive Plasticity*. Cambridge: Polity.
Malm, A. and Hornborg, A. 2014. The geology of mankind? A critique of the Anthropocene narrative. *The Anthropocene Review* 1(1): 62–69.
Manyena, S. 2006. The concept of resilience revisited. *Disasters* 30(4): 433–450.
Manzini, E. 2015. *Design, When Everybody Designs: An Introduction to Design for Social Innovation*. Cambridge, MA: MIT Press.
March, J. 1977. Bounded rationality, ambiguity and the engineering of choice. *The Bell Journal of Economics* 9(2): 587–608.
Margolin, V. 1995. Design history or design studies: subject matter and methods. *Design Issues* 11(1): 4–15.
Marres, N. 2012. *Material Participation: Technology, the Environment and Everyday Publics*. New York: Palgrave.
Marshall, V. 2013. Aesthetic resilience. In S. Pickett, M. Cadenasso, and B. McGrath (eds), *Resilience in Urban Ecology and Design: Linking Theory and Practice for Sustainable Cities*. Dordrecht: Springer.
Marston, S., Jones, J.P., and Woodward, K. 2005. Human geography without scale. *Transactions of the Institute of British Geographers* 30(4): 416–432.
Martin, L. and Secor, A. 2013. Towards a post-mathematical topology. *Progress in Human Geography* 38(3): 420–438.
Martin, R. 2007. *An Empire of Indifference: American War and the Financial Logic of Risk Management*. Durham, NC: Duke University Press.
Marx, K. 1990. *Capital: Volume 1*. New York: Penguin.
Maskrey, A. 1989. *Disaster Mitigation: A Community Based Approach*. Oxfam Development Guidelines No. 3. Oxford: Oxfam.
Maskrey, A. 1994. Disaster mitigation as a crisis of paradigms: Reconstructing after the Alto Mayo earthquake, Peru. In A. Varley (ed.), *Disasters, Development and Environment*. London: Wiley.
Maskrey, A. 2011. Revisiting community-based disaster management. *Environmental Hazards: Human and Policy Dimensions* 10(1): 42–52.
Mason, K. and Whitehead, M. 2013. Transition urbanism and the contested politics of ethical place making. *Antipode* 44(2): 493–516.
Massey, D. 2005. *For Space*. London: Sage.
Massumi, B. 1992. *A User's Guide to Capitalism and Schizophrenia: Deviations from Deleuze and Guattari*. Cambridge, MA: MIT Press.
Massumi, B. 2002. *Parables for the Virtual: Movement, Affect, Sensation*. Durham, NC: Duke University Press.
Massumi, B. 2005. Fear (the spectrum said). *Positions* 13(1): 31–48.
Massumi, B. 2009. National enterprise emergency: steps toward an ecology of powers. *Theory, Culture & Society* 26(6): 153–185.

Masten, A. 2015. *Ordinary Magic: Resilience in Development*. New York: Guilford.
Mbembe, A. and Roitman, J. 1995. Figures of the subject in times of crisis. *Public Culture* 7: 323–352.
McGuire, T. 1997. The last northern cod. *Journal of Political Ecology* 4: 41–54.
McQuillan, D. 2016. The Anthropocene, resilience and post-colonial computation. *Resilience: International Policies, Practices, Discourses*. Pre-print available online at http://www.tandfonline.com/doi/abs/10.1080/21693293.2016.1240779. Accessed 3 May 2017.
Meeks, B. 2000. *Narratives of Resistance: Jamaica, Trinidad, the Caribbean*. Kingston: University of West Indies Press.
Meng, J.C.S. 2009. Donald Schön, Herbert Simon, and *The Sciences of the Artificial*. *Design Studies* 30: 60–68.
Miller, T., Baird, T., Littlefield, C., Kofinas, G., Chapin, F.S., and Redman, C. 2008. Epistemological pluralism: reorganizing interdisciplinary research. *Ecology and Society* 13(2): 46.
Miller, T., Wiek, A., Sarewitz, D., Robinson, J., Olsson, L., Kriebel, D., and Loorbach, D. 2014. The future of sustainability science: a solutions-oriented research agenda. *Sustainability Science* 9(2): 239–246.
Mirowski, P. 2002. *Machine Dreams: Economics Becomes a Cyborg Science*. Cambridge: Cambridge University Press.
Nadasdy, P. 1999. The politics of TEK: power and the "integration" of knowledge. *Arctic Anthropology* 36(1/2): 1–18.
Nadasdy, P. 2003. *Hunters and Bureaucrats: Power, Knowledge, and Aboriginal–State Relations in the Southwest Yukon*. Vancouver: University of British Columbia Press.
Nadasdy, P. 2007. Adaptive co-management and the gospel of resilience. In D. Armitage, F. Berkes, and N. Doubleday (eds), *Adaptive Co-Management: Collaboration, Learning, and Multilevel Governance*. Vancouver: University of British Columbia Press, pp. 208–227.
Nadasdy, P. 2010. Resilience and truth: a response to Berkes. *MAST* 9(1): 41–45.
Nelson, D., Adger, W.N., and Brown, K. 2007. Adaptation to environmental change: contributions of a resilience framework. *Annual Review of Environment and Resources* 32: 395–419.
Nelson, S. 2014. Resilience and the neoliberal counter-revolution: from ecologies of control to production of the common. *Resilience: International Policies, Practices, Discourses* 2(1): 1–17.
Neocleous, M. 2013. Resisting resilience. *Radical Philosophy* 178: 2–7.
Newell, A. and Simon, H. 1972. *Human Problem Solving*. Englewood Cliffs, NJ: Prentice-Hall.
North, D. 1994. Economic performance through time. *American Economic Review* 84(3): 359–368.
North, D. 2005. *Understanding the Process of Economic Change*. Princeton, NJ: Princeton University Press.
O'Brien, K. 2011. Global environmental change II: from adaptation to deliberate transformation. *Progress in Human Geography* 36(5): 667–676.
O'Keefe, P., Westgate, K., and Wisner, B. 1976. Taking the naturalness out of natural disasters. *Nature* 260: 566–567.
O'Malley, P. 2004. *Risk, Uncertainty, and Government*, London: Glasshouse.
O'Malley, P. 2010. Resilient subjects: uncertainty, warfare, and liberalism. *Economy and Society* 39(4): 488–509.

Ó Tuathail, G. 1996. *Critical Geopolitics*. Minneapolis: University of Minnesota Press.
ODPEM 1999. *National Hazard Mitigation Policy: Draft Policy*. Kingston, Jamaica: Office of Disaster Preparedness and Emergency Management.
Odum, N. 1969. The strategy of ecosystem development. *Science* 164(3877): 262–270.
Olsson, P., Folke, C., and Hahn, T. 2004. Social-ecological transformation for ecosystem management: the development of adaptive co-management of a wetland landscape in southern Sweden. *Ecology and Society* 9(4): 2.
Olsson, P., Gunderson L., Carpenter, S., Ryan, P., Lebel, L., Folke, C., and Holling, C.S. 2006. Shooting the rapids: navigating transitions to adaptive governance of social-ecological systems. *Ecology and Society* 11(1): 18.
Olsson, P., Galaz, V., and Boonstra, W. 2014. Sustainability transformations: a resilience perspective. *Ecology and Society* 19(4): 1.
Olsson, L., Jerneck, A., Thoren, H., Persson, J., O'Byrne, D. 2015. Why resilience is unappealing to social science: theoretical and empirical investigations of the scientific use of resilience. *Science Advances* 1(4): 1–11.
Ostrom, E. 1990. *Governing the Commons: The Evolution of Institutions for Collective Action*. Cambridge: Cambridge University Press.
Ostrom, E. 1999. Coping with the tragedies of the commons. *Annual Review of Political Science* 2: 493–535.
Ostrom, E. 2007. A diagnostic approach for going beyond panaceas. *PNAS* 104(39): 15187–15187.
Ostrom, E. 2008. Institutions and the environment. *Economic Affairs*. September 2008: 24–31.
Ostrom, E. 2009. Beyond markets and states: polycentric governance of complex systems. Nobel Prize Lecture, 8 December 2009.
Ostrom, E., Janssen, M., and Anderies, M. 2007. Going beyond panaceas. *PNAS* 104(39): 15176–15178.
Ostrom, V. 1976. Some paradoxes for planners: human knowledge and its limitations. In A. Chickering (ed.), *The Politics of Planning: A Review and Critique of Centralized Economic Planning*. San Francisco: Institute for Comparative Studies.
Ostrom, V. 1980. Artisanship and artifact. *Public Administration Review* 40(4): 309–317.
Ostrom, V. 1997. *The Meaning of Democracy and the Vulnerability of Democracies*. Ann Arbor: University of Michigan Press.
Paine, R. 1966. Food web complexity and species diversity. *The American Naturalist* 100(910): 65–75.
Paine, R. 1969. A note on trophic complexity and community stability. *The American Naturalist* 103(929): 91–93.
Pasquino, P. 1978. Theatrum politicum: the genealogy of capital – police and the state of prosperity. *Ideology and Consciousness* 4: 41–53.
Pasquino, P. 1993. Political theory of war and peace: Foucault and the history of modern political theory. *Economy and Society* 22(1): 77–88.
Peck, J. and Tickell, A. 2002. Neoliberalizing space. *Antipode* 34(3): 380–404.
Peet, R. and Watts, M. 1993. *Liberation Ecologies: Environment, Development, Social Movements*. London: Routledge.
Pelling, M. 1998. Participation, social capital and vulnerability to urban flooding in Guyana. *Journal of International Development* 10: 203–226.

Pelling, M. 2003. Paradigms of risk. In M. Pelling (ed.), *Natural Disasters and Development in a Globalizing World*. London: Routledge.
Pelling, M. 2010. *Adaptation to Climate Change: From Resilience to Transformation*. London: Routledge.
Peterson, G. 2000. Scaling ecological dynamics: self-organization, hierarchical structure, and ecological resilience. *Climatic Change* 44: 291–309.
Philo, C. 1992. Foucault's geography. *Environment and Planning D: Society and Space* 10: 137–161.
Philo, C. 2007. "Bellicose history" and "local discursivities": an archaeological reading of Michel Foucault's *Society Must be Defended*. In J. Crampton and S. Elden (eds), *Space, Knowledge and Power: Foucault and Geography*. Burlington: Ashgate, pp. 341–367.
Pickett, S. and Cadenasso, M. 1995. Landscape ecology: spatial heterogeneity in ecological systems. *Science* 269(5222): 331–334.
Pickett, S. and White, P. (eds) 1985. *The Ecology of Natural Disturbance and Patch Dynamics*. Orlando, FL: Academic Press.
Pickett, S., Cadenasso, M., and Grove, J. 2004. Resilient cities: meaning, models, and metaphor for integrating the ecological, socio-economic, and planning realms. *Landscape and Urban Planning* 69: 369–384.
Povinelli, E. 2006. *The Empire of Love: Towards a Theory of Intimacy, Genealogy, and Carnality*. Durham, NC: Duke University Press.
Povinelli, E. 2011a. *Economies of Abandonment: Social Belonging and Endurance in Late Liberalism*. Durham, NC: Duke University Press.
Povinelli, E. 2011b. Routes/Worlds. *e-flux journal* 27.
Povinelli, E. 2016. *Geontologies: A Requiem for Late Liberalism*. Durham, NC: Duke University Press.
Pugh, J. and Grove, K. 2017. Assemblage, transversality and participation in the neoliberal university. *Environment and Planning D: Society and Space*. Pre-print online version DOI: 10.1177/0263775817709478
Pupavac, V. 2001. Therapeutic governance: psycho-social intervention and trauma risk management. *Disasters* 25(4): 358–372.
Rancière, J. 1995. *Disagreement: Politics and Philosophy*. Minneapolis: University of Minnesota Press.
Rancière, J. 2001. Ten Theses on Politics. *Theory & Event* 5(3).
Rancière, J. 2004. *The Politics of Aesthetics*. London: Bloomsbury.
Rancière, J. 2009. *Aesthetics and its Discontents*. Malden: Polity Press.
Rancière, J. 2013. *Aisthesis: Scenes from the Aesthetic Regime of Art*. London: Verso.
Rip, A. and Kemp, R. 1998. Technological change. In: S. Rayner and E. Malone (eds), *Human Choice and Climate Change, Vol. 2*. Columbus: Battelle Press.
Rist, L., Campbell, B.M., and Frost, P. 2013. Adaptive management; where are we now? *Environmental Conservation* 40: 5–18.
Rittel, H. and Webber, M. 1973. Dilemmas in a general theory of planning. *Policy Sciences* 4: 155–169.
Robbins, P. 2004. *Political Ecology*. Oxford: Blackwell.
Roberts, P. 2013. *Disasters and the American State: How Politicians, Bureaucrats, and the Public Prepare for the Unexpected*. Cambridge: Cambridge University Press.
Rocheleau, D., Thomas-Slayter, B., and Wangari, E. (eds) 1996. *Feminist Political Ecology: Global Issues and Local Experiences*. New York: Routledge.
Rogers, P. 2012. *Resilience and the City: Change, Disorder and Disaster*. Burlington: Ashgate.

Rogers, P. 2013. Rethinking resilience: articulating community and the UK riots. *Politics* 33(4): 322–333.
Rosenblueth, A., Wiener, N., and Bigelow, J. 1943. Behavior, purpose, teleology. *Philosophy of Science* 10(1): 18–24.
Rowe, P. 1987. *Design Thinking*. Cambridge, MA: The MIT Press.
Rüedi Ray, K. 2010. *Bauhaus Dream-House: Modernity and Globalization*. New York: Routledge.
Scheper-Hughes, N. 1992. *Death Without Weeping: The Violence of Everyday Life in Brazil*. Berkeley: University of California Press.
Scheper-Hughes, N. 2008. A talent for life: reflections on human vulnerability and resilience. *Ethnos* 73(1): 25–56.
Schmidt, J. 2013. The empirical falsity of the human subject: new materialism, climate change and the shared critique of artifice. *Resilience: International Policies, Practices and Discourses* 1(3): 174–192.
Schmidt, J. 2015. Intuitively neoliberal? Towards a critical understanding of resilience governance. *European Journal of International Relations* 21(2): 402–426.
Schön, D. 1983. *The Reflective Practitioner: How Professionals Think in Action*. New York: Basic Books.
Schuldenfrei, R. 2009. The irreproducibility of the Bauhaus object. In J. Saletnik and R. Schuldenfrei (eds), *Bauhaus Construct: Fashioning Identity, Discourse and Modernism*. New York: Routledge.
Scott, A. 1988. *New Industrial Spaces: Flexible Production, Organisation and Regional Development in North America and Western Europe*. London: Pion.
Scott, J. 1998. *Seeing Like a State: How Certain Schemes to Improve the Human Condition Have Failed*. New Haven, CT: Yale University Press.
Scott, D. 2003. Political rationalities of the Jamaican modern. *Small Axe* 14: 1–22.
Sent, E.M. 1997. Sargent versus Simon: bounded rationality unbound. *Cambridge Journal of Economics* 21(3): 323–338.
Sent, E.M. 2000. Herbert A. Simon as a cyborg scientist. *Perspectives on Science* 8(4): 380–406.
Sent, E.M. 2001. Sent simulating Simon simulating scientists. *Studies in the History and Philosophy of Science* 32(3): 479–500.
Sent, E.M. 2005. Simplifying Herbert Simon. *History of Political Economy* 372(10): 227–232.
Simon, H. 1955. A behavioral model of rational choice. *Quarterly Journal of Economics* 69(1): 99–118.
Simon, H. 1956. Rational choice and the structure of the environment. *Psychological Review* 63(2): 129–138.
Simon, H. 1957. Application of servomechanism theory to production control. In H. Simon, *Models of Man: Social and Rational*. New York: John Wiley & Sons, pp. 219–240.
Simon, H. 1962. The architecture of complexity. *Proceedings of the American Philosophical Society* 106(6): 467–482.
Simon, H. 1969. Some strategic considerations in the construction of social science models. In P. Lazarsfeld (ed.), *Mathematical Thinking in the Social Sciences*. Glencoe, IL: Free Press, pp. 387–415.
Simon, H. 1972a. The organization of complex systems. In H. Pattee (ed.), *Hierarchy Theory: The Challenge of Complex Systems*. New York: George Braziller, pp. 3–27.

Simon, H. 1972b. Theories of bounded rationality. In C.B. McGuire and R. Radner (eds), *Decision and Organization*. Amsterdam: North-Holland, pp. 161–176.
Simon, H. 1991. *Models of My Life*. New York: Basic.
Simon, H. 1996. *The Sciences of the Artificial* (3rd ed.). Cambridge, MA: MIT Press.
Simon, H. 1997. *Administrative Behavior* (4th ed.). New York: The Free Press.
Simon, S. and Randalls, S. (2016). Geography, ontological politics and the resilient future. *Dialogues in Human Geography* 6(1), 3–18.
Simondon, G. 1992. The genesis of the individual. In J. Crary and S. Kwinter (eds), *Incorporations 6*. New York: Zone Books.
Sloterdijk, P. 2009. *Terror from the Air*. Los Angeles: Semiotext(e).
Smith, A. and Raven, R. 2012. What is protective space? Reconsidering niches in transitions to sustainability. *Research Policy* 41(6): 1025–1036.
Smith, A. and Stirling, A. 2010. The politics of social-ecological resilience and sustainable socio-technical transitions. *Ecology and Society* 15, no. 1 (2010): 11.
Smith, A., Stirling, A., and Berkhout, F. 2005. The governance of sustainable socio-technical transitions. *Research Policy* 34 (2005): 1491–1510.
Smith, N. 1984. *Uneven Development: Nature, Capital, and the Production of Space*. Athens: University of Georgia Press.
Sneddon, C. 2000. 'Sustainability' in ecological economics, ecology, and livelihoods: a review. *Progress in Human Geography* 24(4): 521–549.
Soja, E. 2011. *Postmodern Geographies: The Reassertion of Space in Critical Social Theory* (2nd ed.). New York: Verso.
Star, S. and Griesemer, J. 1989. Institutional ecology, 'translations' and boundary objects: amateurs and professionals in Berkeley's Museum of Vertebrate Zoology, 1907–39. *Social Studies of Science* 19(3): 387–420.
Stoler, A. 1994. *Race and the Education of Desire: Foucault's* History of Sexuality *and the Colonial Order of Things*. Durham, NC: Duke University Press.
Stone, C. 1983. *Democracy and Clientelism in Jamaica*. New Brunswick, NJ: Transaction Books.
Storper, M. 1997. *The Regional World: Territorial Development in a Global Economy*. London: Guilford.
Taussig, M. 1991. *The Nervous System*. New York: Routledge.
Thomas, D. 2011. *Exceptional Violence: Embodied Citizenship in Transnational Jamaica*. Durham, NC: Duke University Press.
Thrift, N. 2004. Intensities of feeling: towards a spatial politics of affect. *Geografiska Annaler B* 86: 57–78.
Tierney, K. 2014. *The Social Roots of Risk: Producing Disasters, Promoting Resilience*. Stanford, CA: Stanford University Press.
Tlostanova, M. 2017. On decolonizing design. *Design Philosophy Papers* 15(1): 51–61.
Tompkins, E., Adger, W.N., and Brown, K. 2002. Institutional networks for inclusive coastal management in Trinidad and Tobago. *Environment and Planning A* 34(6): 1095–1111.
Tonkinwise, C. 2015. Design for transitions – from and to what? *Design Philosophy Papers* 13(1): 85–92.
Tsing, A. 2015. *The Mushroom at the End of the World*. Princeton, NJ: Princeton University Press.
Turnpenny, J. and O'Riordan, T. 2007. Putting sustainability science to work: assisting the East of England region to respond to the challenges of climate change. *Transactions of the Institute of British Geographers* 32: 102–105.

UN. 1989. International decade for natural disaster reduction. General Assembly Resolution A/44/236. New York: United Nations.
UN. 2005. *Hyogo Framework for Action 2005–2015: Building the Resilience of Nations and Communities to Disasters*. Geneva: United Nations.
UN. 2015. *Sendai Framework for Disaster Risk Reduction*. Geneva: United Nations.
Vale, L. and Campanella, T. 2005. *The Resilient City: How Modern Cities Recover from Disaster.* Oxford: Oxford University Press.
Viveiros de Castro, E. 2014. *Cannibal Metaphysics*. Minneapolis: Univocal.
Voß, J., and Bornemann, B. 2011. The politics of reflexive governance: challenges for designing adaptive management and transition management. *Ecology and Society* 16(2): 9.
Walker, B., Carpenter, S., Anderies, J., Abel, N., Cumming, G., Janssen, M., Lebel, L., Norberg, J., Peterson, G., and Pritchard, R. 2002. Resilience management in social-ecological systems: a working hypothesis for a participatory approach. *Conservation Ecology* 6(1): 14.
Walker, B., Holling, C.S., Carpenter, S.R., and Kinzig, A. 2004. Resilience, adaptability and transformability in social-ecological systems. *Ecology and Society* 9: 5.
Walker, B., Gunderson, L., Kinzig, A., Folke, C., Carpenter, S., and Schultz, L. 2006. A handful of heuristics and some propositions for understanding resilience in social-ecological systems. *Ecology and Society* 11(1): 13.
Walker, B. and Salt, D. 2006. *Resilience Thinking: Sustaining Ecosystems and People in a Changing World.* Washington, DC: Island Press.
Walker, B. and Salt, D. 2012. *Resilience Practice: Building Capacity to Absorb Disturbance and Maintain Function*. Washington, DC: Island Press.
Walker, J. and Cooper, M. 2011. Genealogies of resilience: from systems ecology to the political economy of crisis adaptation. *Security Dialogue* 42(2): 143–160.
Walker, R.B.J. 1993. *Inside/Outside: International Relations as Political Theory*. Cambridge: Cambridge University Press.
Walsh-Dilley, M. and Wolford, W. 2015. (Un)defining resilience: subjective understandings of 'resilience' from the field. *Resilience: International Policies, Practices, Discourses* 3(3): 173–182.
Walters, C. 1986. *Adaptive Management of Renewable Resources*. Caldwell, NJ: Blackburn Press.
Walters, C. 1997. Challenges in adaptive management of riparian coastal ecosystems. *Ecology and Society* 1(2): 1.
Walters, C. and Hilborn, R. 1978. Ecological optimization and adaptive management. *Annual Review of Ecology and Systematics* 9: 157–188.
Wamsler, C. 2009. Reducing risk in a changing climate: changing paradigms toward urban pro-poor adaptation. *Open House International* 35(1): 6–25.
Wark, M. 2004. *A Hacker Manifesto*. Cambridge, MA: Harvard University Press.
Washburn, A. 2015. *The Nature of Urban Design: A New York Perspective on Resilience*. Washington, DC: Island Press.
Watson-Verran, H. and Turnbull, D. 1995. Science and other indigenous knowledge systems. In S. Jasanoff, G. Markle, J. Petersen, and T. Pinch (eds), *Handbook of Science and Technology Studies*. London: Sage.
Watts, M. 1983. *Silent Violence: Food, Famine, and Peasantry in Northern Nigeria.* Berkeley: University of California Press.
Watts, M. 2015. Now and then: the origins of political ecology and the rebirth of adaptation as a form of thought. In G. Bridge, J. McCarthy, and T. Perreault (eds), *The Routledge Handbook of Political Ecology*. London: Routledge, pp. 19–50.

Weber, C. 2012. Design, translation, citizenship: reflections on the virtual (de) territorialization of the US-Mexico border. *Environment and Planning D: Society and Space* 30: 482–496.
Weber, C. and Lacy, M. 2011. Securing by design. *Review of International Studies* 37: 1021–1043.
Weichselgartner, J. and Kelman, I. 2014. Geographies of resilience: challenges and opportunities of a descriptive concept. *Progress in Human Geography* 39(3): 249–267.
Welsh, M. 2014. Resilience and responsibility: governing uncertainty in a complex world. *The Geographic Journal* 180(1): 15–26.
Westley, F. 1995. Governing design: the management of social systems and ecosystems management. In L. Gunderson, C.S. Holling, and S. Light (eds), *Barriers and Bridges to the Renewal of Ecosystems and Institutions*. New York: Columbia University Press, pp. 391–427.
Westley, F., Carpenter, S., Brock, W., Holling, C.S., and Gunderson, L. 2002. Why systems of people and nature are not just social and ecological systems. In *Panarchy: Understanding Transformations in Human and Natural Systems*. Washington, DC: Island Press, pp. 103–120.
Whatmore, S. 2002. *Hybrid Geographies: Natures, Cultures, Spaces*. London: Sage.
Whatmore, S. 2013. Earthly powers and affective environments: an ontological politics of flood risk. *Theory, Culture and Society* 30(7–8): 33–50.
While, A., Jonas, A.E.G., and Gibbs, D. 2010. From sustainable development to carbon control: eco-state restructuring and the politics of urban and regional development. *Transactions of the Institute of British Geographers* 35(1): 76–93.
White, D. 2015a. Metaphors, hybridity, failure and work: a sympathetic appraisal of Transitional Design. *Design Philosophy Papers* 13(1): 39–50.
White, D. 2015b. Critical design and the critical social sciences. *Critical Design/ Critical Futures Articles*. Paper 8.
White, G. 1945. *Human Adjustment to Floods*. Department of Geography Research Paper no. 29. Chicago: The University of Chicago.
Whitehead, M. 2007. *Spaces of Sustainability: Geographical Perspectives on the Sustainable Society*. London: Routledge.
Wilson, G. 2012. Community resilience, globalization, and transitional pathways of decision-making. *Geoforum* 43: 1218–1231.
Wisner, B. 1993. Disaster vulnerability: scale, power and daily life. *GeoJournal* 30(20): 127–140.
Wisner, B., Blaikie, P., Cannon, T., and Davis, I. 2004. *At Risk: Natural Hazards, People's Vulnerability and Disasters*. London: Routledge.
Wolpert, J. 1964. The decision process in spatial context. *Annals of the Association of American Geographers* 54(4): 537–558.
Wolpert, J. 1965. Behavioral aspects of the decision to migrate. *Regional Science* 15(1): 159–169.
Wylant, B. 2010. Design thinking and the question of modernity. *The Design Journal* 13(2): 217–231.
Yamane, A. 2009. Climate change and hazardscapes of Sri Lanka. *Environment and Planning A* 41: 2396–2416.
Ye, Q., Hu, X., and Han, Z. 2017. Developing an ICT-based toolbox for resilient capacity building: challenges, obstacles and approaches. In W. Yan and W. Galloway (eds), *Rethinking Resilience, Adaptation and Transformation in a Time of Change*. New York: Springer.

Yeung, H. 2005. Rethinking relational economic geography. *Transactions of the Institute of British Geographers* 30(1): 37–51.
Zebrowski, C. 2016. *The Value of Resilience: Security Life in the Twenty-First Century.* London: Routledge.
Zellmer, S. and Gunderson, L. 2009. Why resilience may not always be a good thing: lessons in ecosystem restoration from Glen Canyon and the Everglades. *Nebraska Law Review* 87(4): 893–949.
Zimmerer, K. 2015. Understanding agrobiodiversity and the rise of resilience: analytic category, conceptual boundary object of meta-level transition? *Resilience: International Policies, Practices, Discourses* 3(3): 183–198.

Index

100 Resilient Cities, 3

abstraction 163, 223; hacking as 253–254; space as 19, 21, 71–72
adaptability 53, 61–62, 65
adaptation 20, 86–87, 141–143
adaptive cycles 80–86, 148
adaptive governance 116–119, 195, 207, 268–269; *see also* co-management
adaptive human behavior 157–158, 159–160, 161
adaptive management 44–45, 93–94, 106–129, 132–133, 149–151, 175, 196, 210, 270; *see also* adaptive governance; affective engineering; transition(s), management
aerial warfare 59–60
aesthetic, modern, and sensory experience 165–166
affective engineering 191–192, 199–203, 272, 273
affects 191–192, 200
agriculture 74, 96, 100, 126–127
airport design 192
Anderies, J. 123–124
Anderson, Ben 30
Anthropocene 262; world-making 242–251
anthropology 30, 247
aquatic ecosystems 35; *see also* rivers; water; wetlands
art, and science 163, 204–205; *see also* Bauhaus
artifacts 153–154, 176; re-purposing 194
artifice 152, 159, 173–177; *see also* cybernetics; design
artificiality 258
assemblage 252

Banham, Reyner 10
basins of attraction 74–76
Baudrillard, Jean 27, 133–134, 165–166, 170
Bauhaus 10, 27, 163–166, 167, 255, 268
becoming otherwise 205; *see also* potentiality; transformations
behaviorist approach 156, 171–172
Benjamin, Walter 161
Berkes, Fikret 104–105, 120, 187, 216, 217
big data 251
biopolitics 190–191, 201, 215, 223–224; *see also* institutional design
biopower 189–192
boundary object 8, 15–16, 42, 47, 48–49, 174
bounded rationality 11, 14–16, 132, 138–142, 147, 149–151, 152, 176
Bowker, Geoff 168–169
Brand, Fridolin 7, 8, 16, 30, 47, 48–49, 169, 174
Brown, Katrina 103–104
Brown, Wendy 50
Buchanan, James 89, 208–209, 210
Buchanan, Richard 11, 12, 192–193, 211–212, 256–257
"Building Disaster Resilient Communities" project (BDRC) 38

calculative mode of thought 71–72, 162
Cameroon, economic crisis 228–229
Canada, wildlife conservation 65
Canguilhem, Georges 158–159
Cannon, Terry 44–45
capitalism 186, 251, 252–253, 257–258
CCTV 37
centralization 26, 53, 91, 95–101, 118; critique 21, 60, 92–93
Chandler, David 53–54, 60, 69, 92, 93, 133, 152, 172, 251
change 6, 20, 24, 78, 101, 125–126; *see also* adaptation; topology; transformations
choice: constitutional 208–209; individual 171–172, 177, 206–207
Cinner, Joshua 126
Clark, Nigel 183, 242, 244, 245, 272
class relations 185, 186
climate change 40–41, 86–87, 183
collaboration 105, 112–115, 124, 130, 220, 237; place-based 260–261
Collier, Stephen 60, 89–91, 208
colonialism 201
co-management 45, 65, 123–128; *see also* adaptive governance; adaptive management
command and control 67, 69–72, 80, 86–87, 90, 91, 102, 187; pathology of 26, 95–101, 105; and representation 113
common pool resources 102, 117–119, 209–210
community 35, 38, 120; building 43, 195, 197, 199–200
community-based disaster management (CBDM) 197–198, 199, 200–203
comparative case studies 123, 126
compartmentalization 98, 143–144
complex systems science 60
complexity 51, 53–54, 83, 91; and collaboration 112–115, 130; hierarchy 83, 131–132, 135; simple or general 92–93, 102–103; *see also* adaptation; cybernetics; decision-making; hierarchy
computer modeling 114–115; *see also* cybernetics
computers 137, 159–161, 198–199; *see also* simulations
conflict 66, 67, 116, 205–208, 210–211; as feedback 238, 266; *see also* power relations; warfare
conservation 65, 98, 103, 247
constructivist approaches 11, 256
context 192–194, 203–204
contingency 32, 57–58, 183, 185, 242
Cooper, Melinda 53, 91
creative destruction 149
crises/disruptions 36–37, 60, 96–100, 151, 161–163, 195; Cameroon 228–229; and hierarchy 143–144; and identity 221; and indeterminacy 242–244; and interdependence 156; *see also* change; disaster management
critical geographers 21–22
critique 28, 89–130, 105, 220–223, 230–231, 268–271; for cybernetic control 183, 184; designerly 203–205; and experimentation 184–188; as feedback 266, 274; human-environment relations 185–187; modernist design aesthetic 183–184
Cross, Nigel 9–10, 256
Crutzen, Paul 244
cybernetics 19, 27, 133–134, 267–268; aesthetics of 161–168; and control 168–169; definition 135, 156; environmental totality 171–177; regulation 253; research 155–161

death/non-death 61–62, 221
Decision Theater 218–219
decision-making 14, 109, 138–140; *see also* adaptive management; bounded rationality
defuturing 257–258, 263, 272
Deleuze, Gilles 19, 55–56, 231, 234–235, 250, 258
democracy 31, 49, 206–207
Derrida, Jacques 57, 231
Descartes, René 71–72, 142, 162
descriptive object, resilience as 15–16, 30, 47, 48–49
design: artifice as 152; critique 203–205; and geographers 275–276; and institutional fit 106; modernist/Bauhaus 133, 138, 163–166, 167; resilience as 131–181; science 153–155; studies 8–18, 154–155;

transition 254–261, 263; and urban planning 37, 41; *see also* context; cybernetics; institutional design; will to design
design studies 258–259
Dewey, John 152, 206
diagrams 19–21, 136, 238–241
difference/multiplicity 24, 184, 214–220, 231, 233, 238, 262, 263; *see also* diversity; geontopower
Dillon, Michael 182
disaster management 37–40, 42, 196–203; *see also* crises/disruptions; security
disaster studies 5–6
disciplinary techniques 190, 201
disruption *see* crises/disruptions
diversity 96–97, 100, 102, 103
Duffield, Mark 59

ecological systems 15–16, 33, 47; and adaptability 34–35, 53; critique and adaptive management 91–130; hierarchy 146–149; reductionist view 68–72; resilience and inequality 65–66; resilience as nomad science 72–86; topological approach 67–68, 80–85; *see also* cybernetics; environment(al)
economics 14, 90, 132, 137, 208–209, 210; *see also* institutional economics, new; neoliberalism
education 237; disaster management 197–199, 201, 202; disciplinary 190, 201
efficiency 97–98, 100
Elden, Stuart 71–72
embodied experience 235
emergence, problems of 51, 156
emergencies *see* crises/disruptions; disaster management
engineering 1–2, 33–35, 36, 38, 68–72, 73–74; affective 191–192; civil 140–141
Enlightenment 166, 185
environment rationalization *see* command and control
environment(al): crises 96–100; management 67–68, 69, 72–73, 80, 91, 127; movement 79–80; power 195, 196–198; studies 7; totality 134, 165–166, 167, 168, 171–177; uncertainty 59–60; *see also* adaptive management; cybernetics; disaster management; ecological systems; sustainability
equilibrium 72
errors *see* mistakes
Escobar, Arturo 258–259
Esposito, Roberto 201
ethics 19, 26, 28, 32, 48, 50; *see also* experimental ethic
Evans, Brad 53, 61, 242
evolution 145
experimental ethic 11–12, 62, 110–112, 121–123, 151, 184, 196, 247; *see also* pragmatic approach; transformations
explication 162, 165

feedback mechanisms 157–158, 176, 237–238, 266
field of adversity 90–91, 93, 95–105
fire management 35, 95–96, 139–140
fishing industry 72–73, 98–100
flood management 246–247
Florida Everglades 126–127
focus groups 199–200
Folke, Carl 44–45, 96, 104, 111, 116–117, 120–122, 123, 125–126, 128, 151, 152, 169, 187, 207, 212, 217, 220
force relations 56, 59, 66
Foucault, Michel 4, 32, 50, 51–52, 57, 58–59, 71, 89, 184–185, 189–190, 223–224
framing, as partial 182
freedom 31, 189; *see also* choice
Fry, Tony 257–258
functional abstraction 168
functional calculus 166–167
functionality 27, 44–45, 134–135, 163–165, 169–170, 172–173, 204; and design 153–154, 176–177; and hierarchy 143–144

Gallie, W.B. 31
Galloway, Alexander 254
genealogies 18–25, 26–27, 32–33, 50–62, 66, 91
geographers 21–22, 275–276
geographic thought 1–29, 230

geometry 71–72, 74, 77
geontopower 249, 272–273
Gillan, G. 58
global biospheric changes 80, 85
globalization, and topology 232
governance reform 44, 92, 194
grid, metaphor 104–105
Gropius, Walter 163
Grosz Elizabeth 204–205, 231
Guattari, Felix 252, 272
Gunderson, Lance 86, 127

hacking 251–254, 262, 274
Hardin, Garret 118
Harvey, David 71, 185–186
Hayek, Frederich 53, 91, 92–93
hazards *see* crises/disruptions; disaster management
hierarchy 70, 82–83, 131–132, 139–140, 152, 172, 209, 232; and adaptation 143; ecosystems 145–149
Hilborn, R. 109–110, 111, 114, 150
history 55–63; bellicose 58–62, 63; total 58, 59
Holling, C.S. 5, 6, 26, 34, 35, 68–74, 77–86, 91, 106, 145–147, 172, 173–176; adaptive management 106–107, 108, 111, 112–113, 114; on command and control 187; and critique 93–94; on panarchy 131; on reduction of variability 96–97
homo economicus 14, 137, 141
Hyogo Framework for Action (2005) 37, 42–43, 196–197

identity, and coping 221, 271–272; *see also* difference/multiplicity
immunization 201
indigenous peoples: justice 248, 249–250, 258, 261; knowledge 120, 214, 216–218, 219, 238, 240
individualism, liberal 59; *see also* neoliberalism
individualization 190
inequalities, socio-economic 6, 43, 65–66; *see also* indigenous peoples
information: and being 27; exchange 182, 213–214; processing 155, 156–157, 161, 217
innovation spaces 128
institutional design 117–119, 123–124, 130, 133, 175, 187–188, 208, 209–210, 211–213, 215–216, 235–238, 270
institutional economics, new 90, 101, 105–106, 152, 173, 208
institutional fit 101–104, 105–106, 122, 208
insurance 236–237
interdependence/interconnection 60, 156–158, 207
interdisciplinary research 30, 47, 49, 105; *see also* boundary object; collaboration
interiority/exteriority 231
international relations 22, 30
interviews 199–200
invasive species 33, 35
Invisible Committee 251

Jamaica 38, 196–198, 201–203, 218
Jax, Kurt 7, 8, 16, 30, 47, 48–49, 169, 174

Kant, Immanuel 184–185, 243
Keynesian economics 90
knowledge 19; as calculable 71–72, 162; subjugated 57, 65–88, 219–221; and uncertainty 120–121; *see also* collaboration; indigenous peoples; local knowledge

Lakoff, Andrew 60
land rights, indigenous 248, 249
land use, watersheds 74–76
language 51–52, 56
leadership 195; *see also* adaptive management
learning 109, 110–111, 214–215; and bounded rationality 150–151; cybernetic vision of 237–238; social 121–122, 207
Lebel, Louis 117
Lee, Kai 5, 113, 114, 132, 147–148, 149–150, 196, 205–206, 207–208, 210
Lemert, C. 58
Lentzos, Filipa 33–34
Levin, Simon 102
liberal individualism *see* neoliberalism
Limits to Growth thesis 79

Lisbon, Portugal 242–243
local communities, and institutional fit 103–104
local knowledge 44, 121, 198–203, 218, 238, 240; *see also* indigenous peoples
local/global processes 232
Lorimer, Jamie 247
Lovelock, James, "Gaia thesis" 79

managerial practice 90–91, 97–98, 100, 104; *see also* adaptive management
Manchester 37
Manzini, Ezio 257, 260–261
Margolin, Victor 8–9, 10
market rationality 54, 92, 93, 118, 208
Marston, Sally 232, 233
Martin, Lauren 231–232
Marvin, Simon 40–41
Marx, Karl 185, 186
matsutake mushroom trade 248
Mbembe, Achille 228–229
meaning 10, 51–52, 56, 122
Meffe, Gary 69–70, 96–97, 187
memory 219–222
Miami Beach 1–4
military strategies 59–60, 157, 162
mistakes 110–111, 157
models/modeling *see* simulations
modernist design movement *see* Bauhaus
modernity, European 184–185
morphogenesis 78
Müller-Mahn, Detlef 44–45
multiple-equilibrium ecosystems 6, 72–74, 79, 81, 83, 158
mutual learning systems 124

Nadasdy, Paul 65–66, 67, 116, 124–125
Neocleous, Mark 41
neoliberalism 14, 45, 51, 53, 54, 91, 133, 186–187, 262, 269–270; and abandonment 228–229, 249; and CBDM 200–202; critique 89–90, 152; demanding environment of 248–250; market rationality 93; reforms 6–7, 248; and uncertainty 59–60
networks 60, 121–122, 124–125
New York, terrorist attacks (11 September 2001) 38–39
Newell, Allen 141–142
North, Douglass 101

Office of Preparedness and Emergency Management (ODPEM) 198
Olsson, Per 116, 125, 126, 127–128, 187–188, 212, 213
ontological design 258, 260
ontological politics 246
optimization 70–71, 79, 80, 109, 111, 158
Ostrom, Elinor 5, 101–102, 117–119, 132, 209–210, 211–212
Ostrom, Vincent 89, 208, 214–215

panarchy 20, 67, 77, 83–86, 131–132
paradigm shifts 65
patch dynamics 82, 83
perspectives 11–12, 17
Philo, Chris 57
place-making 260–261, 276
placements 11, 192–193, 211–212, 256–257
policing 48
policy design process 109–110, 112–113, 114–115, 222
political ecology 186–187
political science 92, 246
politics 48, 188–189, 274; *see also* neoliberalism; power relations
pollution 45, 79, 95
populations 79, 190
positivism 90, 97, 102–103, 105, 113
postmodernism 36
potentiality 56–61, 136, 171, 177, 242, 250, 271–273
poverty *see* inequalities, socio-economic
Povinelli, Elizabeth 244, 248–250, 272–273
power relations 46, 50, 182–183; design out 44–45, 188–189; and sensory experience 165; and truth claims 58–59, 185; *see also* biopower; capitalism; conflict
pragmatic approach 15, 16–18, 150, 204, 206, 207, 229; *see also* experimental ethic
predation 67–68, 72–73
privatization 118
problem-solving 11–12, 135–136

property rights 101, 106, 118
psychology 33
public choice theory 101, 206, 208–211, 212, 214

radical design 257–261, 263
Rancière, Jacques 48, 165–166
Randalls, Samuel 32
rationality 14, 132, 137–138, 143, 166; and technology 162–163; *see also* abstraction; bounded rationality
recovery 68
reductionism 68–69, 161
redundancy 100, 121, 235
reflection 247, 256
regulation 46, 168–169, 189, 196; see also security
Reid, Julian 53, 61, 182, 242
representation, and prediction 113–114
resilience, definition 169
resistance 68
resource extraction 66
resource management 80, 95–100, 118–119, 120–121, 187, 208, 217; *see also* fishing industry
revolution 163, 185
re-wilding 247
risk 41–42, 190–191, 269
Rittel, Horst 11
rivers 43
Rockefeller Foundation 3
Roitman, Janet 228–229
Rose, Nikolas 33–34
Rosenblueth, Arturo 156, 157
Roussel, Raymond 52

Salt, David 61, 182–183, 221
Sanderson, Geoff 106, 131
scalar thinking 232
Schawinsky, Alexander 164
Schmidt, Jessica 152
Schön, Donald 255–256
scientific knowledge: and art 204–205; limits of 17, 130; and management 112–115; research 91
sea levels 2
Secor, Anna 231–232
security 22, 36–37, 41, 46, 59–60, 156, 189–191, 266; *see also* disaster management
self-help 41, 43
Sendai Framework 37
sense-making 12, 234–235
sensory experience, as contingent 165–166
serialization 234–235
servomechanism theory 157–158
Simon, Herbert 4–5, 9, 10–11, 14–16, 83, 108, 112, 131–161, 171–173, 254–255; *Administrative Behavior* 139–141; *Models of My Life* 177; *Sciences of the Artificial* 152, 153–155, 177
Simon, Stephanie 32
simulations 113–115, 148, 159–161, 167, 170–171, 198–199, 218–219
Sloterdijk, Peter 162
slowness 246–247
social barriers 213–214
social control 171–172
social innovation, design for 260–261, 263
social learning 121–122, 207
social memory 104, 219–221
social service cuts 6, 45
social systems 96, 100, 116, 121–122
socio-ecological traps 126
solutions-oriented design 22, 24
solutions-oriented research 17–18
sovereignty 31
spaces of dispersion 56
space/spatial analysis: abstraction 19, 21, 71–72; genealogy as 51–55; *see also* topology
species diversity 146–147
stability/instability 73, 74, 85
standardization 99, 100
state space 74–77
statistics 70, 190–191
Stockholm Conference (1972) 79–80
storm events 39–40
structuralism 231–232
subjectivity, and truth 167
subjugated knowledge 57, 65–88, 219–221
sublime, the 243
surveillance 46
sustainability 85–87, 149–150, 206, 235–236, 239, 258
synthesis, functional, and design 165, 167–168, 223

synthetic reasoning 17–18, 22, 27
systems theory 17, 109–110, 111, 113–114, 144–145, 160; systemic dynamics 147–148; *see also* hacking

technology: and art 163; and rationality 162–163
Tierney, Kathleen 5–6
time, as change and becoming 233
time-space 84–85
topology 23–24, 67, 90, 230, 231–234, 263, 267; ecological resilience 80–85; as state space 74–77
totalization 185, 204
traditional knowledge 216–217; *see also* indigenous peoples
transformations 116, 117, 120, 125–126, 188–189, 194–195, 212; *see also* conflict; experimental ethic
transgression 24–25, 89, 91, 184, 185
transition(s): design 254–261, 263; management 46–47, 128
Transitions Towns Movement 258, 259
transversality 252–253, 272
traps, resilience 126–127
trauma 33, 39
trust and engagement 35
truth 19, 58–59, 166–167, 185, 204; *see also* will to truth

uncertainty 34–36, 59–60, 111, 120–121, 230
un-grounding 272
urban resilience 1–4, 16, 36–37, 39–41
urban security 36–37

variability *see* diversity
vector field, for meaning 238–241
violence, industrialized 162
von Neumann, John 137
vulnerability 6, 186, 240, 269; causes and self-help 41, 42–46

Walker, Brian 61, 182–183, 221
Walker, Jeremy 53, 91
Walters, Carl 108–109, 110, 111, 114, 115–116, 150
warfare 57, 59–60, 162–163
Wark, McKenzie 251, 252
water: agriculture 127; management 209; urban resilience 1–2
watersheds 74–76
Weiner, Norbert 157–158
Westley, Frances 105, 215, 217
wetlands 126–127
Whatmore, Sarah 246–247
wicked problems 11–12
wildlife conservation 65, 98, 103, 247
will to design 4–5, 8–25, 27, 168, 175–177, 222, 241, 273; *see also* design
will to truth 13–14, 21–22, 135, 177, 221; *see also* truth
windows of opportunity 195
Wylant, Barry 12, 193, 203–204

the thing about Media Consumption is that it feeds us our representations — it interprets the world for us.

- by giving us representations, we not only lose our creative abilities, we are also unwittingly driven towards the goals embedded in those representations.

1) Connection between representation and pragmatics/goals/values
2) sources/mechanisms of representation, how we absorb/release it (Media)
3) What is resistance/freedom? Can we ever create our own vision

Made in the USA
Monee, IL
21 October 2020

45777387R00177